CONTESTING INTERSEX

Contesting Intersex

The Dubious Diagnosis

Georgiann Davis

NEW YORK UNIVERSITY PRESS

New York and London

NEW YORK UNIVERSITY PRESS
New York and London
www.nyupress.org

References to Internet websites (URLs) were accurate at the time of writing. Neither the author nor New York University Press is responsible for URLs that may have expired or changed since the manuscript was prepared.

Library of Congress Cataloging-in-Publication Data
Davis, Georgiann.
Contesting intersex : the dubious diagnosis / Georgiann Davis.
pages cm   Includes bibliographical references and index.
ISBN 978-1-4798-1415-2 (cloth : alk. paper) — ISBN 978-1-4798-8704-0 (pbk : alk. paper)
1. Intersex people. 2. Intersexuality—History. 3. Sexual disorders. I. Title.
HQ78.D39   2015
306.76'85—dc23   2015009606

New York University Press books are printed on acid-free paper, and their binding materials are chosen for strength and durability. We strive to use environmentally responsible suppliers and materials to the greatest extent possible in publishing our books.

Manufactured in the United States of America

10 9 8 7 6 5 4 3 2 1

Also available as an ebook

To Barbara for her unfailing support from the beginning

And to Katrina for lending me strength and teaching me resiliency

CONTENTS

ACKNOWLEDGMENTS

I can't believe I've written a book about intersex. Only ten years ago, I was too ashamed of my intersex trait to say the word *intersex*, let alone spend years of my life writing about it. Words cannot do justice to the complexities of this transformation, but they can be used to thank the many people from the intersex community who showed me the possibilities inherent in bridging personal experience and professional passion. I'm fortunate to follow the path blazed by Jim Ambrose, Claudia Astorino, Eden Atwood, Arlene Baratz, Max Beck, Janik Bastien-Charlebois, Tony Briffa, Mauro Cabral, Peggy Cadet, David Cameron, Morgan Carpenter, Caitlin Petrakis Childs, Hiker Chiu, Mo Cortez, Cary Gabriel Costello, Martha Coventry, Katie Baratz Dalke, Tiger Devore, Betsy Driver, Jane Goto (who gave me the shirt off of her back, literally), Janet Green, Thea Hillman, Curtis Hinkle, Morgan Holmes, Leslie Jaye, Cynthia Johnson, Amber Jones, Emi Koyama, Bo Laurent, Lynnell Stephani Long, Amy Lossie, Mani Mitchell, Angela Moreno, Iain Morland, Sherri Groveman Morris, Miriam Muscarella, Jeanne Nollman, Pidgeon Pagonis, Emily Quinn, Daléa Rundblad, Galen Sanderlin, Kimberly Saviano, Lianne Simon, Margaret Simmonds, Anne Tamar-Mattis, Sugee Tamar-Mattis, Kiira Triea, Daniela Truffer, Miriam van der Have, Hida Viloria, Sean Saifa Wall, Karen Walsh, Gina Wilson, Kimberly Zieselman, and countless others. Without these trailblazers, many of whom I discovered in the crucial volume *Intersex in the Age of Ethics* (Dreger 1999), I would never have had the idea to write this book. And without the generous community members who were willing to share their experiences with me, I would not have been able to blaze my own trail, for both the book and the transformative personal journey that accompanied it.

I collected the powerful tools for the sociological toolbox with which I constructed this book from my former professors, especially David Asma, Mike Armato, Bill Bielby, Sharon Collins, Cary Gabriel Costello, Rohan de Silva, Nilda Flores-Gonzalez, Chava Frankfort-Nachmias, Lorena Garcia, Ron Glick, Christina Gomez, Sydney Hart, Jennifer Jordan, Andreas Savas Kourvetaris, Anthony Lemelle, Pat Martin, Nancy Mathiowetz, Wamucii Njogu, Barbara Risman, Laurie Schaffner, BarBara Scott, Susan Stall, Brett Stockdill, Martha Thompson, and Steve Warner.

When I was a doctoral student at the University of Illinois at Chicago, Barbara Risman mentored me in many ways, from reading countless drafts of my work to challenging me to think far outside my comfort zone. She helped me put together a dream team committee, including Verta Taylor from the University of California, Santa Barbara; Sharon Preves from Hamline University; and Lorena Garcia and Laurie Schaffner from the University of Illinois at Chicago (UIC), whose insights got this project on its way. Barbara also did everything she could to make sure I had the funding I needed to collect and analyze my data, including, at UIC, the Al and Betty Brauner and Scott Brauner Fellowship, the Kellogg Merit Scholarship, the Rue Bucher Memorial Award, and a Chancellor's Research Fellowship, as well as the 2009 Beth B. Hess Memorial Scholarship, awarded jointly by Sociologists for Women in Society and the Society for the Study of Social Problems, with additional support from the American Sociological Association. Most important, Barbara believed in me, a working-class student with a very nontraditional educational history. I learned much from her about the profession, life, friendship, and what it's like to be mentored by a feminist. She changed my life, and I hope to model her mentorship with my own students for years to come.

I still find it both comic and tragic that I received my first real information about my own intersex trait, not to mention the existence of an intersex community, from sociocultural scholars, not from the doctors who surgically altered my body. The words of so many, including Claudia Astorino, Janik Bastien-Charlebois, Monica Casper, Cary Gabriel Costello, Alice Dreger, Anne Fausto-Sterling, Ellen Feder, Emily Grabham, Morgan Holmes, Rebecca Jordan-Young, Annemarie Jutel, Katrina Karkazis, Suzanne Kessler, Geertje Mak, Iain Morland, Sharon Preves, Lizzie Reis, Katrina Roen, David Rubin, Margaret Simmonds, Alyson Spurgas, Brian Still, and Sarah Topp, continue to change my life and influence my work. I know that Erin Bergner and Rodney Hunt will join this illustrious list.

I'd like to thank Jill Bystydzienski, David Hutson, Lih-Mei Liao, PJ McGann, Barbara Risman, Katrina Roen, Virginia Rutter, Mindy Stombler, Jennifer Suchland, Rebecca Wanzo, Elroi Windsor, and anonymous reviewers at various journals for their feedback on my earlier work. I'm also grateful to audience members at various sociological meetings, intersex conferences, continuing medical education events, and university brown bags and symposia where I've presented my work over the years for their comments and questions that have substantially enhanced my thinking. I would never have arrived at the theoretical and empirical analysis presented here without their feedback. A special thanks to Susan Stryker for organizing and inviting me to present at the University of Arizona's James J. Leos Symposium "Intersex: Medical, Cultural, and Historical Contexts," where I had incredibly fruitful conversations with other panelists and participants that stayed with me as I finished this book.

I've benefited from a great deal of technical assistance at all stages of this project. Ilene Kalish and her editorial team at New York University Press saw promise in this book and used their sharp editorial skills to make it stronger. During data collection, Jen de la Rosa, Amy Kalov, Amanda Marquez, Michael Pijan, and Stephen Terrazas were instrumental in ways I can't even begin to quantify. Kaylin James, Amanda LaGrow, and Paige McLeod provided useful feedback during the earlier stages of the writing process. Annemarie Jutel, Sharon Preves, Lizzie Reis, and an anonymous reviewer offered helpful comments as I shaped the final manuscript. In the last inning, Rebecca Steinitz came to bat with meticulous line edits and sharp comments, and Lisa DeBoer offered her skills in the construction of the index.

My colleagues at Southern Illinois University Edwardsville (SIUE) and the University of Nevada, Las Vegas (UNLV), encouraged me at different points along the way. At SIUE, where I began the book, I'd especially like to thank Kevin Cannon, Teri Gulledge, Erin Heil, Flo Maätita, Linda Markowitz, Sarah Morrison, Erin Murphy, Laurie Puchner, and Catherine Seltzer. The process of finishing it at UNLV helped me feel at home in my new university. I'd particularly like to thank Christie Batson, Michael Ian Borer, Barb Brents, Lynn Comella, Robert Futrell, Jennifer Keene, Ranita Ray, and Anna Smedley for their assistance during my transition to UNLV.

Through various chapters of my life, I've looked up to and been inspired by Lisa Berube, Cati Connell, Kiana Cox, Ilana Demantas, Jodie

Dewey, Erika Gray, Maggie Hagerman, Mosi Ifatunji, Laura Logan, Sarah Lynch, Michelle Manno, Jennifer Maple, Juan Martinez, Erin Murphy, Zachary Neal, Jennifer Prange-Hauptman, and Marco Roc.

My family—especially my parents, Georgia and George; my brother, Nick; my grandparents Anna and Nick; and my aunts, Nancy and JoAnne—has always helped me as best they could. I will always unconditionally love them, and I recognize that they are always in my corner.

Finally, I'd like to thank my friends, who have kept me grounded and emotionally stable through some of the choppiest waters of my life. This book wouldn't exist without the support of Rachel Allison, Amy Brainer, Courtney Carter, Marcia Klink, Linda Markowitz, Sharon Preves, Ranita Ray, Barbara Risman, and Pratim Sengupta. A very special final shout-out to Pallavi Banerjee, Katrina Karkazis, Jennifer Kontny, and last, but not least, the two- and four-legged animals who fill my home. I'm glad to be sailing the seas of life with these friends by my side.

1

Introduction

"You're in the Monkey Cage with Me"

Summer in Chicago is a time for exploring the city, visiting Lake Michigan, enjoying great food, and debating the causes of the Cubs' inability to win. However, in the summer of 2008, a group of more than one hundred visitors chose to stay in their run-of-the-mill conference hotel rather than discover what Chicago has to offer. Mostly white middle-class women, young and old, from around the country, they had come to the Windy City for the annual meeting of the Androgen Insensitivity Syndrome Support Group–USA (AISSG-USA).

Unlike most other national conferences, this one had no posters with its name decorating the hotel. Instead, the signs that led attendees to their registration site displayed a colorful orchid and the words "Women's Support Group" in bold letters, followed by "Please do not disturb." Curious hotel employees and guests found the meeting mysterious. One employee asked me, "Why are all of you needing support?" I shrugged my shoulders, not sure how to respond.

In fact, the ambiguity of the signs was appropriate to the purpose of the meeting, whose attendees were distinguished by the fact that the sex they were born with had been deemed biologically ambiguous by the medical profession. To put it another way, almost everyone at the meeting had been born with an intersex trait (or accompanied someone born with an intersex trait). In many instances, the result was physical bodies incongruent with sex chromosomes.[1] In the past, these individuals might have been considered hermaphrodites, a term that some—but not all—in the intersex community now consider derogatory. Terms less contentious today include *intersex*, *intersex traits*, *intersexuality*, and *intersexual*, which I will use throughout *Contesting Intersex*.

This book is about how *intersex* is defined, experienced, and contested in contemporary U.S. society.[2] I argue that medical professionals have replaced intersex language with disorder of sex development nomenclature, a linguistic move designed to reclaim their authority and jurisdiction over the intersex body. Ironically, this disorder of sex development (DSD) terminology was strategically introduced in 2005 by Cheryl Chase, a prominent intersex activist, and her allies, who had hoped the new nomenclature would improve medical care for those born with intersex traits.[3] Instead, as I show here, DSD terminology has heightened tension within the intersex community. Some individuals born with intersex traits are embracing the new nomenclature; others resist it, citing the pathologization that underlies the term *disorder*; a few are indifferent to diagnostic labels and think individuals should use whatever term(s) they prefer.[4] My hope is that *Contesting Intersex* will tease apart the tensions over terminology in the intersex community, while also showing how power resides in diagnostic labels.

Although *intersex* is itself a term whose meaning is contested, in general it is used to describe the state of being born with a combination of characteristics (e.g., genital, gonadal, and/or chromosomal) that are typically presumed to be exclusively male or female. People with androgen insensitivity syndrome (AIS), for example, have XY chromosomes and testes "but lack a key androgen receptor"[5] that consequently prevents their bodies from responding during gestational development and beyond to the normal amounts of androgens (an umbrella term for testosterone) their testes produce. Depending upon how much androgen the receptor blocks, some AIS individuals have ambiguous external genitalia (usually a larger clitoris that resembles a small penis) with either internal or external testes, while others have an outwardly "normal"-looking vagina with a shortened vaginal canal, no uterus, and undescended testes. Swyer Syndrome, sometimes referred to as XY gonadal dysgenesis, is another example of an intersex trait. Like individuals with AIS, people with Swyer Syndrome have sex chromosomes that are inconsistent with their phenotype (external appearance). Swyer Syndrome is characterized by the presence of testes (usually internal), but unlike those with AIS, individuals with Swyer Syndrome usually have a uterus, though it is generally smaller than usual. People born with traits such as AIS or Swyer Syndrome are sometimes referred to as "intersex," meaning they do not fall into the binary of "male" or "female." According to the Intersex Society of North America (ISNA), there are approximately twenty different intersex traits.[6]

There is no simple medical explanation for the cause(s) of intersex, no agreement on what defines intersex, and no formal record of those born with such "abnormalities."[7] All of these lacks presumably contribute to the challenge of establishing the frequency of intersex. Still, estimates have been made, with the most-used figure suggesting 1 in 2,000 people is intersex,[8] but because estimates drastically vary across publications,[9] I'm uncomfortable offering my own estimate. What I do know, however, is that intersex people exist all around the world.

Estimates of intersex in the population did not matter to the conference attendees, who shared a unique medical history and had a strong connection with one another. They were there to support one another in healing from what, for many, has been a life full of medical lies, deception, and unnecessary surgical intervention. Yet if you had happened to stumble into that Chicago hotel that summer weekend, you would have had no idea of what had brought the attendees together. Without the "Women's Support Group" signs, you might have imagined that you were interrupting a meeting of sorority sisters or a family reunion. A group official told me that one reason for the secrecy was to prevent any attendees from feeling uncomfortable or "freakish." This was also why the support group's public website did not name the conference location.

I found AISSG-USA through that very website, as I searched the Internet for information about *my* intersex "abnormality." As a twenty-seven-year-old individual with complete androgen insensitivity syndrome (CAIS), I had met only one other intersex person, a friend from work who was as private about her diagnosis as I was about mine. I wanted to know more about intersex, and I started my search for information online. Although I did not know it at the time, in doing so I was employing what Nikolas Rose and Carlos Novas (2005) label digital biocitizenship. To Rose and Novas, digital biocitizenship connects one electronically to a specific community via, for example, a support group website or e-mail listserv.

I was diagnosed with CAIS around the age of thirteen. I was experiencing abdominal pains, and my mother thought I would soon begin menstruating, a rite of passage for women in my family, as in many other families. However, my period never came. The abdominal pain went away, but my mother was concerned enough to seek medical advice. I soon found myself in an endocrinologist's office, wondering why so many doctors were literally looking over—and within—my body. At the time, the doctors told me I had underdeveloped ovaries that had a very high risk of being cancerous and would need to be surgically removed before my eighteenth birthday. But the doctors were lying: The purported ovaries were actually undescended testes. Encouraged by medical providers, my parents went along with the lie, and when I was seventeen, I had surgery to remove the supposedly dangerous organs.

I would not see a doctor again, or discover that I had an intersex trait, until, at nineteen, I relocated to a new area far from my childhood medical providers, where I sought new doctors. As is customary, they requested that I bring my medical records with me to my appointment for a routine physical. When I finally got my hands on my surgical records, I read them in utter disbelief. That was my first encounter with the truth about my body and the medically unnecessary surgery I had undergone. At that time, it made me deeply uncomfortable to learn that I had XY chromosomes and testicular feminization syndrome—the label for my trait when I was initially diagnosed. I was in tears as I read what one gynecologist had written in my medical file: "After extensive discussion I feel pt [patient] needs surgery to have gonads removed. She is not aware of any chromosomal studies and most literature agrees it best she not be aware of the chromosome studies. She has been told she is missing her uterus, she does have a vagina. She has no tubes. She has been told she may have streaked ovaries and they should be removed because of the possibility of developing gonadal cancer" (personal medical records, November 26, 1997).

I was shocked and confused. Why had my medical providers and parents lied to me for so many years? I thought I'd had surgery because of a health risk.[10] Was having an intersex trait that horrible? I remember thinking I must be a real freak if even my parents hadn't been able to tell me the truth. I ran to the dumpster outside my building and threw my records away, not wanting to be reminded of the diagnosis or the surgery that couldn't be undone.

Almost a decade later, I was finally emotionally ready to confront my medical past, and I requested another set of my records. I was exploring feminist theories and gender and sexuality scholarship in my sociology doctoral program, an incredibly empowering experience that positioned me to revisit my personal experience with sex, gender, and sexuality binaries. Our assigned readings and thoughtful classroom discussions encouraged me to delve deeply into my medical history, first with close graduate school friends and faculty, eventually with anyone who cared to listen. Finally feeling liberated, I sought others like me, which is how I ended up at my first intersex support group conference, which happened to be in Chicago that year.

This project was born during that emotional weekend, which will forever mark my first involvement with the intersex community. Although I entered the weekend as an individual with an intersex trait looking for peers, by its end I was determined to pursue a sociological analysis of intersex in contemporary U.S. society. Despite my personal experience with intersex, I was initially concerned that I would have difficulty gaining access to a community that hasn't had the best experiences with researchers, notably psychologist John Money, whose work was discredited after the discovery of his falsified data and unethical research practices (see chapter 3).[11] I was wrong: AISSG-USA was incredibly supportive of my research, as was each of the other organizations I studied. It became clear during data collection that my personal experience with intersexuality provided an inroad into the field and community that would eventually become my second home (see Figure 1.1). In the fall of 2013, I was elected president of the AIS-DSD Support Group, the new name of AISSG-USA, which had just started to allow men with intersex traits to attend their annual meeting (clearly, a lot had changed since I'd attended my first meeting).

As a result of the access, support, and assistance I received in connecting with intersex community members, I was able to collect a tremendous amount of data in a relatively short time. During this period, I formed friendships throughout the intersex community, to which I am now permanently connected. For I am not only studying the intersex community, I'm in it. As Peggy, a fifty-six-year-old with an intersex trait, said to me, "I feel that you're going to be on my side. You're not like someone at the zoo saying, 'Well I'm a human being and I'm taking notes on the monkeys.' You're in the monkey cage with me."

Like that of many of my research participants, my experience with intersexuality has left me with some horrific physical and emotional scars. However, it has also become the core of my intellectual passion and academic commitment. On January 7, 2010, in the midst of my data collec-

Figure 1.1. A publicly available photo of individuals with intersex traits at the 2010 Annual Meeting of AISSG-USA. I am surrounded in this picture by a handful of AISSG-USA members who, like me, have an intersex trait. I am the one standing in the middle of the crowd wearing a dark button-down and glasses. Photo uploaded to Wikipedia by Ksaviano.

tion, I met fifty-three-year-old Cheryl Chase, who was instrumental in the rise of the intersex rights movement nearly two decades earlier. As we finished our emotionally intense interview, Chase wrote on a piece of paper, "Georgiann, Finish your PhD and change the world!" That note, along with a framed picture of the two of us, greets me every time I sit down at my desk. It is not just memorabilia from the field; it is symbolic of my commitment to *our* entire community, no matter how divided—over medical terminology and how best to advocate for change—we are today.

The remainder of this chapter introduces the theoretical and methodological underpinnings of my study. I begin with a discussion of gender structure theory, a framework I rely on to understand the complexities of intersex in contemporary U.S. society. I then turn to my efforts to develop trust with the community before collecting interview data, the process of participant recruitment, and an overview of my participants. I end this section by turning the methodological lens on myself. While I acknowledge that my insider/outsider status offered access and insight into the intersex community, I use the concept of the looking-glass self to explain how I unintentionally altered my appearance throughout data collection to match how I *believed* I would be perceived by those I was studying, regardless of whether such perception was warranted. This raises the concern that researchers may self-police their gender presentation. Studying the intersex community, I have been able to combine my personal and professional interests in an attempt to understand the complexities of living outside sex, gender, and sexuality binaries. My hope is that this understanding will help to elucidate how those binaries constrain all of us, whatever our genitalia.

Applying Gender Structure Theory

Contesting Intersex builds on the short history of sociocultural studies of intersex. For example, in 2003, sociologist Sharon Preves published *Intersex and Identity*, wherein she argues that intersex people confront through a series of overlapping stages of identity formation the stigma of being differently bodied. Preves documents how intersex support groups provide a venue for intersex people to connect with one another, resulting in the identity-based intersex rights movement. Preves's work was groundbreaking, but subsequent events have turned it into social history. In 2008, medical anthropologist and bioethicist Katrina Karkazis introduced an impressive methodological approach to the study of intersex in her book *Fixing Sex*. She was the first scholar to bring together rigorously the perspectives of multiple stakeholders in the community: intersex people, their parents, and medical experts. Karkazis eloquently shows how the medicalization of intersex—and the ideologies of sex, gender, and sexuality that are at the core of that medicalization—perpetuate medical authority over the intersex body. Her focus on multiple stakeholders and ideology makes her work an important precursor to *Contesting Intersex*, which moves forward to focus on intersex after the formal introduction of DSD terminology. But my concerns also differ in two important theoretical ways. First, I consider how the social construction of diagnoses made it possible for *intersex* to be transformed into a *DSD*, and second, I ground my analysis of intersex in gender structure theory.

Intersex is a problem because it disrupts the traditional gender order. If our behaviors weren't constrained by gender, if opportunities weren't

filtered through gender, and if gender weren't tied to bodies and identities, it is doubtful that intersex would be as problematic throughout the
world as it is today. Thus, I ground this study of intersex in gender theory,
specifically in the idea that gender needs to be understood as a stratification system residing not only in individuals but also at the institutional
and interactional levels of society (e.g., Butler 2004, 1993, [2006]1990;
Martin 2004; Risman 2004, 1998; Ferree, Hess, and Lorber 1999; Lorber
1994; Connell 1987). To do so, I frame my analysis with sociologist Barbara Risman's gender structure theory (2004, 1998), which I find particularly useful because it can be adapted to integrate other theoretical
insights, such as the ideas of sociologists Nikolas Rose (2007, 2001) and
Carlos Novas (Rose and Novas 2005) about the simultaneously individualizing and collectivizing nature of biological citizenship, Annemarie
Jutel's (2011, 2009) theoretical contribution about diagnoses, and philosopher Giorgio Agamben's (2005, 2000) account of state of exception.

Gender structure theory conceptualizes gender as "deeply embedded
as a basis for stratification not just in our personalities, our cultural rules,
or institutions but in all these, and in complicated ways" (Risman 2004,
433). The institutional dimension of gender structure dictates—through
a variety of structural processes, including organizational practices and
policies about behaviors (e.g., Giddens 1984)—a powerful conceptualization of gender that is tied to bodies. The individual level of gender structure is where the development of a gendered self emerges through the
internalization of a male or female identity and its assigned personality
attributes (e.g., Chodorow 1978). Within this individual level of gender
structure, we come to see ourselves as male or female and masculine or
feminine—as if doing so were that simple, achievable, and unchanging.
Although we see one another's genitals only in very intimate settings, we
constantly make assumptions about whether one is male or female on
the basis of gendered clothing, mannerisms, and hairstyles. This is the
interactional dimension of gender structure and it is where people "do
gender": a function of social psychology that involves categorization by
sex, which triggers stereotypes and gender expectations that ultimately
influence how a person is treated, approached, and expected to behave
in social circumstances (West and Zimmerman 1987; see also Garfinkel
1967 and Goffman 1976, 1959). Each of the dimensions of gender structure shapes and is shaped by the others. For example, our internalization

of a gender identity occurs in relation to how the law defines us at the institutional level and how, at the interactional level, we are expected to present ourselves and are consequently perceived by others.

The explanation in *Contesting Intersex* of how *intersex* became a *DSD* pays specific attention to the consequences of meanings of intersex at the institutional, individual, and interactional levels. At the institutional level, I explore these questions: How does intersex activism disrupt medical authority? How does the medical establishment make sense of and approach intersex bodies? What are the current treatment protocols, and what do these protocols imply about the medical institution's power and control over bodies? At the individual level, I focus on how individuals make sense of their diagnosis and understand the new DSD nomenclature, asking: What benefits and what challenges arise for those who reject DSD terminology? How are these terminological preferences related to one's understanding of gender as either an essentialist characteristic of the body or a socially constructed phenomenon? At the interactional level, I ask: How do medical professionals present intersex to parents of newly diagnosed intersex children? Does the way in which intersex is presented to parents influence how they respond to medical recommendations? What happens when parents learn about intersex from the intersex community, specifically from intersex adults and from other similarly situated parents? How do parents feel about the new DSD nomenclature? All of these levels are essential to my analysis.

Trust Matters

My process of building relationships with key players in the intersex community began when I attended my first AISSG-USA meeting in the summer of 2008, before I started to conceptualize my research. I attended the meeting as an individual with complete androgen insensitivity syndrome, not a researcher. I participated in workshops, including an intensely emotional session in which we shared our most painful stories about intersexuality. I explained how I had been compelled to revisit my intersex diagnosis and tackle it head-on by reading in a doctoral seminar on feminist theory about intersexuality and intersex advocacy. I sat in the classroom thinking: *Where* is this intersex rights movement? My personal experience was so different from what I was reading that my professional interest was piqued. I started reading scholarly work on intersexuality. With each new piece, I found greater liberation. I discovered I wasn't abnormal. As I shared my story, others listened with compassion. They smiled, passed tissues, and offered chocolate. During that private session, as I saw how intersex was still a source of shame, secrecy, and pain for so many in my new-found community, I realized that my sociological study of intersex was needed.

At my very first AISSG-USA meeting, I connected with organizational board members, intersex adults, parents of intersex children, and medical experts on intersexuality. Although I immediately started planning my study and received institutional review board approval in October 2008, I didn't conduct my first interview until the summer of 2009. I believed from early on that building a rapport with key participants was necessary to the research process—especially in this instance, where participants may have had problematic experiences with other researchers. I wanted them to know they could trust me and the broader research process. According to sociologists William Foote Whyte (1984) and Ann Oakley (1981), interview data is only as good as the trust present during the actual interview. More specifically, Whyte identifies rapport as the first concern of the interviewer. I would go further and argue that it is the most important part of an interview. Without rapport, why would participants want to share their information and knowledge with a researcher?

Having entered the community as an individual with an intersex trait, I was exposed to the contemporary struggles in unique ways. My access reached further than traditional participant ethnography, because my first experiences with the community were truly participatory. I started wearing my researcher hat only after my first AISSG-USA meeting. This unconventional inroad into the field allowed me to formulate questions directly from the most natural observations possible. In turn, my methodological approach relied on the trust afforded to me by my own lived experience with intersexuality.

Participant Recruitment

My original intent was to interview up to one hundred individuals affiliated with the intersex community—or to conduct interviews up to the point where I was no longer collecting new information, which is commonly referred to as saturation. I selected participants based on current or past involvement with ISNA, AISSG-USA, Organisation Intersex International (OII), and Accord Alliance. I chose these four organizations because their websites suggested that they were involved in intersex advocacy in different ways. In chapter 2, I discuss each organization's origins and development, including its goals and mobilization strategies. Here, I focus on the participants I interviewed, including intersex people, parents of intersex children, medical experts on intersex, and organizational board members.

I had no trouble recruiting participants for my study. When word got around that I was a researcher who had an intersex trait, I was contacted by many individuals who not only were willing to be interviewed but who also expressed a sincere desire to participate. Because ISNA, AISSG-USA, OII, and Accord Alliance are organizations with activism components, I anticipated that some participants would be eager to discuss their activism and organizational affiliations, and indeed they were. I also had little difficulty recruiting medical professionals who self-identify as experts on intersex. Many of the medical professionals I invited to participate agreed to do so on the basis of reciprocity. Medical professionals routinely ask individuals with bodies like mine to participate in their research studies, and, as one doctor politely put it when at an intersex support group meeting I handed her an invitation to participate in my research, "I'm always asking you people to help me, so of course I will help you."

In total, I interviewed 65 individuals, including 36 adults with intersex traits; 17 parents of children with such traits; 10 medical experts on intersex,[12] including surgeons, urologists, endocrinologists, and mental health professionals; and 2 social movement organizational board members who are not medical professionals, do not have an intersex trait, and are not the parents of an intersex child. (Appendix A includes a list of participant pseudonyms and select demographic characteristics.) Interviews ranged from 25 minutes to well over three hours. I traveled all over the United States to conduct interviews face-to-face in order to gain participants' trust and establish a level of comfort possible only in person. Participants chose the interview locations, which ranged from a semi-private area in a hotel during a support group meeting to homes

ethnic minorities at large experience intersexuality in similar ways to white individuals. In other words, I'm suggesting that the cultural capital gained from educational attainment may explain, at least partially, the lack of any observed relationship between race/ethnicity and experience with intersexuality. This is, of course, another empirical question that is worthy of further investigation.

Most of my participants presented as men or women rather than say genderqueer, and my interactional classifications were consistent with how they referred to themselves.[15] I found classifying their sexuality to be much more difficult. The sexuality I assumed, based on cues tied to gender presentation, didn't always match with a participant's self-reported sexuality. For example, several people I assumed were lesbian by how they dressed and/or wore their hair actually identified as heterosexual. This observation is not surprising, as we know we cannot, nor should we try to, predict one's sexual identity by one's gender presentation.

Insider/Outsider

I knew from the start that I would not study intersexuality for academic purposes exclusively. As an individual with an intersex trait, I am personally connected to the intersex community and advocacy movement whether I choose to be or not. I am thus committed to presenting a scholarly study of intersexuality from the perspective of an intersex feminist academic. As feminist intersex advocate Emi Koyama and interdisciplinary scholar Lisa Weasel documented in 2002, most of what we know about intersexuality is the work of academics and clinicians who do not have a personal connection to intersex; while such perspectives can provide valuable contributions, they do not make up the entire puzzle.

Since Koyama and Weasel's critique of intersex scholarship by non-intersex academics and clinicians, several academics with intersex traits have publicly written about intersex. In 2008, Morgan Holmes, a Canadian sociologist, drew on her personal experience with intersex in her compelling account of intersex adults and their struggles in a world that continues to construct sex and gender as natural binary characteristics. Holmes specifically problematizes intersex medical care by highlighting the complicated relationships between binary ideologies about the body and biomedical power and authority. As an "inside informant," she has produced work that is priceless (Holmes 2008, 64). For more than a decade, Iain Morland, a cultural critic in the United Kingdom, has published extensively on intersex, including his 2009 "scrutiny of intersex activism" wherein he examines the complexities "between activism's critique of medicine and medical reform" (191). Cary Gabriel Costello, a sociologist at the University of Wisconsin–Milwaukee, has also written extensively about intersex from the perspective of an intersex person who has gender transitioned. He maintains two public blogs—The Intersex Roadshow[16] and TransFusion[17]—on which he discusses an array of topics, from intersex fertility to sex/gender checkboxes that we are forced into on forms such as birth certificates.

Like Holmes, Morland, and Costello, among others, I integrate my personal experience and my professional analysis in *Contesting Intersex*. However, my specific goal is to understand how *intersex* became a controversial *DSD*. My unique insider/outsider access to the intersex community enabled me to collect data that might otherwise have remained unavailable to such a project. Individuals can hold an insider or outsider standpoint in every community; the production of scientific knowledge is best served by collecting perspectives from each and all of these standpoints.[18] As much as we are all insiders within certain communities, we are also outsiders to different communities. I define *insider* in the research context as a researcher who has a personal and historical connection to the population or phenomenon being studied. My definition of *outsider* in the research context is a researcher who does not have a personal connection to the population or phenomenon being studied.

I occupied a unique insider/outsider status throughout most of this project. I was an insider in the sense that I was born with an intersex trait. My participants and I shared experiences of living with an intersex trait. However, when I began this project, I was not truly an insider, for I had no historical connection to the intersex community or even to any other individuals with intersex traits. In fact, it wasn't until the 2007–8 academic year that I was fully aware an intersex community even existed. As a result, I define insider/outsider status in the research context as a researcher who has a personal connection with the phenomenon being studied but no historical connection.

My insider/outsider status proved to be incredibly useful not only in gaining access to the intersex community but also in the type of data I was able to gather. Because I am an intersex person, participants on all sides of the community expressed increased comfort in speaking with me. For example, Ann, a fifty-two-year-old with an intersex trait, said, "It made it more comfortable . . . it's easier for me to talk to you." Because I experienced living with intersex, I was personally familiar with many topics that surfaced in interviews, from terminology to difficulty of engaging in penetrative sex because of a shortened vaginal canal and/or vaginal dryness. Jenna, a thirty-one-year-old with an intersex trait, described how my personal familiarity made interviews go more smoothly: "I guess as far as ease of explanation [goes] it's a little easier because you know the terminology. You know when I say TFS . . . you know I meant testicular feminization . . . and when I say AIS or I say gonadectomy or I say whatever . . . you know what I'm talking about. You know it, you have it." My personal experience with intersexuality also made some people more willing to participate in my study. As Marty, a forty-three-year-old parent of a teenager with an intersex trait, put it: "You've experienced this . . . you have a connection . . . If you were just somebody, or if your child didn't have it, or you didn't have it, then I think I wouldn't do this [*points to recorder*] . . . it's like a sisterhood." While many participants shared Marty's view, others said they would have participated regardless of my personal experience with intersexuality.

Still, given my insider/outsider status, these individuals expressed the type of comfort researchers strive for. Mariela, a twenty-nine-year-old with an intersex trait, summarized that comfort: "Well, I guess I might've said the same things to someone else, be it a doctor or a total stranger, and in the back of my mind I would've been like, 'What do they think about that?' Or 'What, are they trying to make up their own mind about me?' Or how I feel about things, I would worry about what they were thinking." Karen, a fifty-two-year-old with an intersex trait, put it similarly:

Do I feel different talking to you about it? Yeah, of course I do. And in a good way. Because first of all, you are a person with a DSD and so you have a level of sensitivity to this topic that a lot of other people don't, and I think that that's a good thing. I don't translate that as you have a whole lot of preconceived notions about it. Maybe you do, I don't know. But at a certain level, I don't care whether you do. I know that you're asking questions and you're asking them in a way that hasn't been asked before.

Although I don't necessarily see my personal experience as the sole cause of my ability to ask questions "in a way that hasn't been asked before," I do believe that my initial insider/outsider status uniquely positioned me to ask questions to a diverse group of people that other researchers might not have thought about or have access to, given their outsider status. This enabled the collection of more complete information.

Turning the Methodological Lens

In 1902, sociologist Charles Horton Cooley introduced the concept of the "looking-glass self." In short, this concept suggests that individuals see and consequently construct themselves in relation to how they believe they are perceived. Cooley argues, "Each to each a looking-glass, Reflects the other that doth pass" (152). We can apply the looking-glass self to research methodology, specifically the collection of data. Cooley maintains: "As we see our face, figure, and dress in the glass, and are interested in them because they are ours, and pleased or otherwise with them according as they do or do not answer to what we should like them to be; so in imagination we perceive in another's mind some thought of our appearance, manners, aims, deeds, character, friends, and so on, and are variously affected by it" (152).

Turning the methodological lens on myself, I realized that in all of my field notes, I mentioned what I was wearing and how my hair was styled, despite the fact that, outside of those notes, I never consciously acknowledged that I was altering my presentation for data collection purposes. For example, after one interview I wrote:

Ashley, my hair stylist, straightened my shoulder length dark brown hair late last night. I was worried that my hair would get messed up while sleeping, or if not while sleeping, get messed up on the very early flight out of Chicago this morning, but thankfully it didn't. Note to self: I really can't wait until I can chop off this hair. However, I really am afraid that if I did, parents won't speak to me, especially those who are really homophobic. While short hair isn't necessarily a sign of being gay, I feel [that] my height (5'10") coupled with the fact that I am very overweight will result in [my] being read as a butch lesbian. Can you imagine that? A parent who wants to express to me their biggest, darkest fear about their intersexual child being gay wouldn't be able to . . . or at least wouldn't feel comfortable because, in essence, they would be directly insulting me. It's a good thing I have Ashley to straighten my hair . . . and it's a good thing my body blocks the hormones that would likely result in [my] having the type of hair that needs daily washing . . . . Let's face it, I have hair like my grandmother's that only needs to be washed once a week, if that! I really can't wait to cut my hair . . .

Clearly I was making decisions about my presentation throughout data collection. However, I never really considered this as a theme until I re-read my field notes during data analysis. As I looked back, I realized that I went an entire year without cutting my hair. After having longer hair for more than a year, I simply couldn't continue with it, so I cut it all off during the last summer of data collection. Even after it was short, when I was in the field I regularly found myself styling my hair in a more feminine way than its usual stereotypically butch look.

There is nothing new in the insight that we choose hairstyles based on how we want to be perceived. My point here is not so much about my hair length or style; rather, it is that as a researcher I made decisions about the length and style of my hair given how I believed my participants would perceive me. In this way, I build on sociologist Sara Crawley's point that "gender identity is always dependent on how audiences will allow us to see ourselves. . . . [C]lothing is more than outward performance. It is also inward performance" (2008, 377). When I interviewed people who I knew identified as gay, lesbian, or queer, my self-policing of gender, which I assumed would be a marker of sexuality, was absent. I wore what I wanted to wear, regardless of how I might be perceived, and similarly styled my hair. In this regard, I constructed myself as queer for the queer people I interviewed and as straight/normative for all others. I was, as sociologist Carla Pfeffer (2014) articulates, accomplishing a gender and sexuality; however, later, after considerable reflection, I realized I was subconsciously presenting myself as a straight woman, rather than as a queer person, in certain "field" instances, in an attempt to minimize any number of limitations I might have encountered throughout data collection, given my queer identity.

The existence of the looking-glass self in the methodological arena is what interests me here. While interview effects have long concerned survey methodologists[19] and qualitative researchers[20] alike, their focus seems to overlook how we, as researchers, shift our identities to appease our own assumptions. In some cases, this might mean constructing a visible self that is in direct opposition to the way we believe we will otherwise be perceived. Research shows that by constructing an identity tied to a university, researchers can increase our response rate as a result of the legitimacy implicitly conferred by the academy.[21] Consequently, researchers usually make their institutional affiliation known during data collection. The new element here is that when studying issues of gender and sexuality, researchers may present themselves in ways related to how they assume they will be perceived, not how they actually are being perceived. The implication of this observation is that researchers may be, consciously or not, self-policing their gender presentation.

Organization of This Book

Although methodology is important in any empirical project, it is at the heart of *Contesting Intersex* given my unique position as a researcher and the nature of my data. Thus it has been the primary focus of this introductory chapter.

In chapter 2, I explore the institutional level of gender structure by focusing on key social movement organizations in the intersex rights movement and the transformation of intersex rights advocacy from "collective confrontation" to "contested collaboration." Intersex social movement organizations (ISMOs) are especially important because, as I will show, they can serve as vehicles for influencing public ideologies about intersex and ultimately changing medical care. My analysis begins with a discussion of the ISMOs I studied (and from which I recruited research participants): ISNA, Accord Alliance, AISSG-USA, and OII. The early years of intersex activism, I argue, involved individuals with intersex traits coming together in a display of what Nikolas Rose and Carlos Novas (2005) define as a rights biocitizenship: "activism such as campaigning for better treatment, ending stigma, gaining access to services, and the like" (442). This particular employment of a rights biocitizenship centered on a collective challenge to the medical profession's approach to intersexuality; thus I label this period "collective confrontation." I also make the case that throughout—and beyond—this period of intersex activism, electronic media were critically important in connecting people across the intersex community. This digital link is what Rose and Novas conceptualize as a digital biocitizenship. Digital biocitizenship enabled individuals in the intersex community to learn about intersex from one another in their own space, rather than in a space created and controlled by the medical community, which resulted in what Rose and Novas call informational biocitizenship: highly specialized knowledge about one's trait. I next describe the emergence of DSD language and the community conflict about terminology that followed; this conflict marks the current period, which I label "contested collaboration." Today's intersex advocacy reflects disagreements and disputes exacerbated by DSD language, as some members of the community want to work with medical professionals to change intersex medical care, while others resist such collaboration. Although the intersex community and its social movement organizations were never monolithic,[22] I show that the introduction of DSD nomenclature heightened differences across the community and the ISMOs in ways that further divided individuals with intersex traits.

Chapter 3 continues to address the institutional level of gender structure but focuses specifically on the medical management of intersex traits. I argue that insights from the sociology of diagnosis—specifically Annemarie Jutel's (2011, 2009) theoretical insights about diagnostic naming—can help us understand the contemporary medical management of intersex, specifically why and how medical experts so quickly embraced DSD language. Today, it is rare to find a medical expert who speaks or writes about intersex without using DSD terminology. Chapter 3 tells the story of how this circumstance came to be. It begins with a discussion of John Money's model for treating intersex, which dominated intersex medical care for much of the second half of the twentieth century. Although Money was once a highly regarded psychologist at The Johns Hopkins Hospital, I revisit the events that led to the discrediting of his research and the collapse of his reputation as the leading expert on intersex. Against the background of this important history, I show how contemporary U.S. medical experts on intersex, like much of society, tend to hold narrow, essentialist understandings of sex, gender, and sexuality. The danger in medical professionals' holding these views is that they often are used to justify medically unnecessary and irreversible surgical interventions on intersex bodies, which have long-lasting implications for intersex people. Taking into account these two realities alongside 1990s intersex activism (discussed in chapter 2), I argue that medical professionals took so quickly to the new DSD nomenclature because it allowed them to escape their tainted history of intersex medical care. With medical authority and jurisdiction over the intersex body in jeopardy, the new language allowed medical professionals to reassert their power over intersex. Medical professionals no longer treat intersex traits; they treat DSD and in doing so maintain their authority.

In chapter 4, I highlight the individual level of gender structure. Rose and Novas's (2005) discussion of biological citizenship again comes into play, but here I focus on biological citizenship at the level of the self. This shift is possible because Rose and Novas conceptualize biological citizenship as "both individualizing and collectivizing" (2005, 441). Biological citizenship at the individual level is particularly concerned with how individuals use biomedical language to describe aspects of the self. I contend that, for intersex people, biological citizenship at this level is accessible only to those willing to engage with DSD nomenclature (see also Rose 2007, 2001).

My argument about the relationship between DSD terminology and access to biological citizenship begins with a discussion of the emotional and sexual struggles faced by individuals with intersex. While these struggles might seem to be the consequence of medically unnecessary surgical interventions, those who were not surgically modified reported similar difficulties. This leads me to conclude that the struggles associated with intersex do not derive exclusively from the scalpel but rather reside within the broader medicalization[23] process, including the pathologization of the intersex body. For this reason, I fear that DSD terminology, if not approached carefully, is potentially dangerous to the intersex community. *Disorder of sex development* implies that one has an abnormality or, worse, is abnormal.

Because *DSD* was formally introduced by the powerful institution of medicine, the intersex community has no choice but to engage with it.

Those who embrace DSD language tend to hold more essentialist understandings of gender—that is, to see gender as biologically prescribed. Embracing DSD terminology puts an individual in a position to access biological citizenship and, most important, its benefits, including support from medical providers and family members; but this access comes at a cost, notably anxiety about feeling abnormal, that DSD language inherently perpetuates. Those who hold more socially constructed views of gender tend to reject DSD nomenclature on the grounds that its pathologizing terminology conflicts with how they understand their intersex bodies and ability to "do gender."[24] Those who dislike DSD language may still identify as male or female, presenting themselves with masculine or feminine cues such as clothing, hairstyles, and the like, but because they don't see themselves as having a *disorder*, they seem to have difficulty accessing biological citizenship and the benefits it can afford. Without access to biological citizenship, individuals are left with minimal, if any, support from medical providers and family members. On a positive note, people who reject DSD language tend to describe more positive senses of self. Because they don't feel that their intersex trait is an abnormality, they do not feel abnormal.

Those who neither embrace nor reject DSD nomenclature give us important insight into the workings of terminological preferences. By being flexible with terminology, people position themselves to have more access to biological citizenship—and its benefits—while also holding a positive intersex identity. In other words, people can selectively employ whichever terminology they believe will be most beneficial at any given time. If they desire support from medical providers and/or family members, they can adopt the medicalized DSD nomenclature. If they begin to feel abnormal, they can embrace intersex language and the notion that intersex is a natural variation of the body. However, I maintain that this flexible access to biological citizenship necessitates an understanding that sex, gender, and sexuality, as well as diagnoses, are socially constructed phenomena.

In chapter 5, I turn to the interactional level of gender structure to investigate how parents of intersex children reach decisions about medical intervention. The chapter begins by looking at how the medical profession presents the intersex trait to parents and how parents respond. I argue that medical professionals construct the intersex trait as a medical emergency that must be addressed immediately and in a medical setting. This creates what Giorgio Agamben (2005, 2000) describes as a state of exception. Medical professionals who frame intersex as an emergency are creating a state of exception that allows them to abandon medical ethics that warn against performing medically unnecessary surgery on children. Once the intersex trait is presented as an emergency and the state of exception is established, medical providers tend to inundate parents with information about intersex. However, the information they present focuses on the alignment of sex, gender, and sexuality as essentialist characteristics of the body, laying the groundwork for justifying medically unnecessary interventions, notably irreversible surgical procedures that many doctors continue to, even today, recommend without any hesitation to parents of newly diagnosed children.

Relying once again on Rose and Novas (2005; see also Rose 2007, 2001), I show how medical professionals then place the responsibility for medical decisions entirely on parents, thereby avoiding responsibility for questionable interventions. Most important, I argue that when parents are exposed to a different kind of information—that is, information which originates in the intersex community rather than in the medical profession—they are more likely to delay or even refuse medical recommendations. Such questioning of medical recommendations directly challenges medical authority and changes the standard course of intersex medical care. Those who obtain information from the intersex community after consenting to medical procedures tend to report a tremendous amount of guilt, but they too challenge medical authority by advising other parents in the intersex community to question the necessity of medical interventions.

Despite their challenges to medical authority, I find that parents of intersex children are not as divided over DSD terminology as intersex adults. Although not all parents embraced DSD nomenclature, many did. While views on DSD language among adults with intersex traits tended to coincide with conceptualizations of gender as either an essentialist characteristic of the body or a socially constructed phenomenon, this pattern was not observed among parents. Their children's intersex traits challenged many parents to adopt more socially constructed views of gender, making conceptualizations of gender less relevant to their positions on nomenclature. Instead, their terminological preferences seemed to have more to do with their acceptance of LGBT communities. Parents who embraced DSD language tended to be critical of the move to include an *I* for *intersex* on the LGBT abbreviation, leading me to believe that homophobia fueled at least some of their terminological preference.

Chapter 6 concludes *Contesting Intersex*. It begins with the public attention intersex has recently received. I focus specifically on a groundbreaking 2013 lawsuit filed in both federal and state courts against "South Carolina Department of Social Services (SCDSS), Greenville Hospital System, Medical University of South Carolina and individual employees" by parents who adopted a child with an intersex trait.[25] I then turn to the possibility for positive social change in the form of practical actions that intersex activists and allies can take to decrease intersex stigma and the shame and secrecy that surround it. These actions include: (1) continuing to fight for the elimination of medically unnecessary surgeries; (2) collaborating with medical allies; (3) forging connections across groups in the intersex rights movement with a goal of increasing gender, racial, and class diversity across and within intersex organizations; (4) overcoming the fear of public exposure; (5) engaging with formal and informal feminist scholarship; (6) recognizing that social constructions—most notably sex, gender, sexuality, and medical diagnoses—drive inequalities in our community; and, most important, (7) valuing the voices of intersex children in the evaluation of intersex medical care.

An Analysis from Within

My personal experience with intersex has substantially shaped my career in unimaginable ways, of which *Contesting Intersex* is the most significant piece of evidence. There is no question that my lived familiarity with intersex has shaped this project from conceptualization to data collection to data analysis. However, throughout each stage, I have also stayed true to feminist methodologies. I supplemented interviews with ethnographic observations; I recognized patterns; and, most important, I questioned data that seemed to deviate. This process involved returning to the data on numerous occasions to revise my typologies. Many times throughout data collection I thought I had found a pattern only to have it disrupted by a particularly informative interview. Following sociological tradition, I continued this process until I was confident the patterns I observed were as stable and predictable as possible. I imagine that some will read this analysis with trepidation or skepticism, given my position in the community, and I welcome such readings. However, by the end of my data collection, I was convinced that I had captured an accurate account of the U.S. intersex community at this point in history. This project has been incredibly invigorating for me personally and professionally, and I hope it similarly invigorates others, whatever their connection to the intersex community.

2

The Transformation of Intersex Advocacy

I first read about intersexuality as a college student studying gender in a sociology classroom, when I was assigned a handful of the classic publications, including Anne Fausto-Sterling's (1993) "The Five Sexes" and excerpts from Suzanne Kessler's (1990) "The Medical Construction of Gender." As evidence that I wasn't some sort of rare freak of nature, these pieces were personally (and later professionally) important to me. I remember thinking that if I was reading about intersex traits in a college classroom, they must be common enough. Still, as an undergraduate, I wasn't ready to disclose my personal experience with intersex. Years later, in the sociology graduate program at the University of Illinois at Chicago, I read Stephanie Turner's article "Intersex Identities" in *Gender & Society*. Turner concludes: "Embodying what they feel is a failure of medicine to make them what they cannot be in the first place, [intersex people] envision a wholly new intersection of sex and gender, a kind of 'third sex' that evades gender determination yet also somehow solidifies into a category of identity" (1999, 458). Reading this, I imagined that intersex people were a radical group of gender rebels, and I just needed to find them.

When I attended my first AISSG-USA meeting, I expected to encounter a strong intersex identity nestled within an embodied health movement[1] focused on contesting the idea that intersex was an abnormality.[2] The little I had read about the intersex rights movement led me to assume it would be a monolithic collective determined to confront medical experts about the ways they treated individuals with intersex traits. This was not the case. Rather than collective confrontation, I found medical providers giving presentations and many intersex people using the term *DSD* casually in conversation, like the woman in her twenties who asked me, "Do you have a DSD?" But there was also a minority who were openly critical of the medical providers in attendance and, more important, the DSD language that was being widely used by providers and other participants alike. These individuals sat together at meals and in sessions, clearly considering themselves to be apart from what appeared to be the mainstream. I found these differences in the intersex community at once surprising, confusing, and intriguing, given what I had read about intersex identity in Turner's article.

I might not have been so perplexed if I had read more widely on the topic, especially the work of interdisciplinary scholar Morgan Holmes, medical anthropologist and bioethicist Katrina Karkazis, and sociologist Sharon Preves. Preves provides a broader view of intersex politics:

> As is true with other social movements, there is a diversity of groups within the intersex movement and not all agree on the goals or methods of making change. In fact, there has been considerable tension between intersex activists about the objectives and tactics of the movement, including how best to frame intersex to be most attractive to potential movement recruits and supporters. As the number of intersex groups, voices, and opinions proliferated, struggles over the methods and very purpose of the movement ensued. The very definition of what it means to be intersexed is politicized, contested, and fraught with conflict, as is the objective of such mobilization. Such tensions are a predictable element of frame alignment processes, indeed because individuals and groups bring competing ideologies to their interactions with one another. (2005, 262)

This account of the intersex rights movement was much more in line with what I was observing, and I decided that Turner's conclusion was no longer an accurate representation of the contemporary intersex community.[3] Indeed, many of the groundbreaking intersex studies produced by sociocultural scholars are now social history, as the intersex community and its advocacy efforts have transformed alongside changes in medical care, most notably the introduction of the new DSD nomenclature.

My analysis of the transformation of intersex advocacy picks up where these influential scholars of intersex left off. In this chapter, I attempt to uncover the complexities of this new era of intersex advocacy by exploring three major questions. First, how did DSD language emerge and spread? Second, why did ISNA, a key U.S. intersex social movement organization, close its doors in the summer of 2008? Third, what mobilizing strategies and tactics do intersex social movement organizations currently employ, and how, if at all, are they related to understandings of gender structure and its institutional, individual, and interactional consequences?

I begin to answer these questions by describing how twentieth-century intersex medical care and feminist critiques thereof sparked the birth of intersex advocacy and its powerful intersex social movement organizations—none of which, it is likely, would have occurred without the expansion of the Internet. I next describe how U.S. intersex advocates initially engaged in collective confrontation against the medical profession in the 1990s, forcing the profession to respond with the 2000 medical statement "Evaluation of the Newborn with Developmental Anomalies of the External Genitalia" (Committee 2000). I go on to show how, despite this response from the medical profession, several key intersex advocates adopted a new political strategy of working with medical professionals that ended up creating new divides in the movement, particularly concerning diagnostic nomenclature.

The Birth of U.S. Intersex Advocacy and Intersex Social Movement Organizations

Since the formation of U.S. intersex social movement organizations in the 1990s, they have played an important role in the intersex community and its advocacy efforts. They bring together people with similar experiences and concerns in ways that position them to promote change at the institutional, individual, and interactional levels of gender structure.[4] The institutional level of gender structure is especially important because it is where institutions such as the medical profession and family formally support and enforce ideas about gender and policies about bodies.[5] It is also where intersex social movement organizations themselves operate and organize for social change by forming biosocial communities[6] that seek to self-educate their members and change intersex medical care, albeit in different ways. The organizations I studied all seek to improve the lives of those born with intersex traits. Yet their strategies for doing so vary according to how they understand the gender structure and conceptualize its relationship to the problems associated with intersex traits. Thus, while members of intersex social movement organizations are positioned to confront the gender structure, that doesn't necessarily mean that they do so.

I focus here on four nonprofit intersex social movement organizations: ISNA, AISSG-USA, OII, and Accord Alliance. I selected these four organizations because my assessment of their publicly available mission and goals, coupled with informal feedback from community members, made it clear that they are—or, in the case of ISNA, were—involved in intersex advocacy in different ways.[7] ISNA was an activist organization that sought to change how medical professionals approached intersex traits. It also served as an informal support group for many intersex people. AISSG-USA has always been primarily a support group, but it also does advocacy work. In recent years, AISSG-USA has coordinated a continuing medical education program in conjunction with its annual meeting. OII is an activist organization that works on reducing the shame and stigma associated with intersex by raising public awareness, and it does not shy away from criticizing the medical profession. Accord Alliance is an organization that promotes change in intersex medical care by working with the medical community through medical education. Although there are other organizations in the movement, these four represent their diversity.

Although some individuals affiliated with these organizations might not describe their involvement in the intersex community as activist in nature, I still refer to them as intersex activists. As I see it, any involvement in an organization with peers who have a common goal to promote change at any level of society—institutional, individual, and/or interactional—is a form of activism. After my first experience at AISSG-USA, it was immediately clear to me that I would need to study multiple intersex social movement organizations if I wanted to capture the variations in experience and strategy among the intersex community. In addition to recruiting interview participants and conducting ethnographic observations in the public meeting spaces of organizational meetings when time allowed, I also analyzed each organization's publicly available material, including their websites, brochures, and handbooks (Consortium 2006a,b), focusing specifically on history, goals, membership, mobilization strategy, and definitions of intersex traits. When it came to recruiting individual participants, I extended beyond these organizations by employing methodological snowball sampling. I asked my initial participants to name others who they believed held different views from their own, which in a few cases led me to people who were not—or were no longer—affiliated with any of the organizations.

The intersex rights movement—and the social movement organizations that fuel it—was formed to change the way the medical profession treats intersex people. Although surgical modification of intersex bodies predates the twentieth century (Mak 2012; Reis 2009; Dreger 1998a), technological advancements across the 1900s, such as the discovery and analysis of sex chromosomes,[8] gave the medical profession more tools for discovering and surgically addressing intersex "abnormalities" (see, for example, Karkazis 2008; Preves 2003, 2001, 2000; Fausto-Sterling 2000a, 1993; Dreger 1998a,b,c; Kessler 1998, 1990; Kessler and McKenna 1978). Trans history is important in understanding the recent history of intersex medical care. The first "sex-change" surgery on a trans person took place in Denmark in the 1920s (Stryker 2008; Meyerowitz 2002; Feinberg 1996; MacKenzie 1994), and in 1952, American Christine Jorgensen made history when she traveled to Copenhagen to undergo a "sex-change" procedure so that her body aligned with her feminine presentation (Jorgensen [1967] 2000). When Jorgensen returned home, she was the feature story in news reports across America, increasing attention to sex reassignment across the U.S. medical profession and reinforcing doctors' authority over bodies, sexualities, and gender. Endocrinologist Harry Benjamin, a prominent figure in trans medical history, had recently "used the term transsexual [in a paper presented at a major medical conference] to describe individuals who felt 'trapped in the wrong body'" (MacKenzie 1994, 44), further enhancing public and medical interest. This public spotlight on surgical modifications of bodies that defy the sex categorization system validated the already-in-place surgical "treatment" of intersex bodies (see Mak 2012; Reis 2009; Dreger 1998a). In other words, technological advancements of the twentieth century, along with media attention on trans issues, made it easier for doctors to define intersex bodies, like trans bodies, as "abnormal" and in need of medical and surgical attention (intersex medical care is discussed in greater detail in chapter 3).

Although surgeries on trans bodies are, in practice, similar to surgeries on intersex bodies, there is a notable difference in agency (Feinberg

1996). Intersex people usually have little, if any, autonomy over the medical management of their bodies, as doctors frequently perform "normalizing" operations on children. Trans people, on the other hand, can decide for themselves whether or not to undergo surgery (provided, of course, that they have access to supportive doctors and can afford the costs associated with such procedures). Feminists who conceptualize sex and gender as socially constructed binaries have long been critical of the medical management of intersexuality (e.g., Kessler and McKenna 1978),[9] especially the irreversible surgical interventions doctors enact on the bodies of healthy intersex babies and young children. In "The Five Sexes: Why Male and Female Are Not Enough," Anne Fausto-Sterling (1993) argued that if we must organize our bodies into sex categories, we ought to move beyond male and female and include true hermaphrodites ("herms"), male pseudohermaphrodites ("merms"), and female pseudohermaphrodites ("ferms").[10] Although she later claimed to be writing with "tongue firmly in cheek,"[11] Fausto-Sterling defined "herms" as having one ovary and one testis, "merms" as having "normal" components of female genitalia but testes instead of ovaries, and "ferms" as having normal components of male genitalia with ovaries instead of testes. By arguing for the recognition of five sexes, Fausto-Sterling attempted to debunk the binary sex system.[12] In 1998, Suzanne Kessler critiqued Fausto-Sterling's taxonomy in her book *Lessons from the Intersexed*, arguing that we need to think in terms of sex "variability" rather than sex "ambiguity." According to Kessler, it is neither possible nor logical to maintain the sex system once we recognize the existence of multidimensional sex variability. To categorize intersex people by sex, she maintains, is to perpetuate the validity of the categorization system. In a piece titled "The Five Sexes, Revisited," Fausto-Sterling accepted Kessler's critique, acknowledging that "It would be better for intersexuals and their supporters to turn everyone's focus away from genitals" (2000b, 22). By providing an analytic ground for activists to challenge the medical profession, these feminist writers helped to spark the intersex rights movement, which sought to change how medical professionals treat intersex.[13]

Fausto-Sterling, for instance, is credited with indirectly facilitating the formation of ISNA, at one time the world's largest intersex advocacy and support group (Preves 2003; Fausto-Sterling 2000a,b; Chase 1998a,

1997). When I interviewed founder Cheryl Chase, she described to me how Fausto-Sterling's scholarship motivated her:

> So I picked up *Myths of Gender* by Fausto-Sterling . . . and it says, "So what is the difference between male and female?" It says well, men are people who impregnate and ejaculate, and women are people who ovulate and menstruate and gestate, and I forget what else. I thought man . . . that is really fucked up. So I called her, at Brown [University], and I said, "In fact, the way that sex differentiation happens, you're not going to be able to draw a clear line between male and female, no matter what you do. Wherever you draw the line, I can produce examples of real-life people who would produce surprising results." She said, "Yes, I agree with you. I've changed my mind since I wrote that and in fact I have this article in press right now, and the article is 'The Five Sexes.'" And I said, "Let me see it." She sent it to me, and I said, "Well, part of what you missed here is how horrible it is what they do to people like me now, and partly, this thing 'The Five Sexes' . . . I hate it, it's stupid. It's not any less stupid than two sexes. It ratifies gonadal histology, which nobody can even see. What kind of thing is that?" She said, "I'll tell the editors [of *The Sciences*, where "The Fives Sexes" will be published that] they should [also] publish a letter from [you]." So I wrote a letter to *The Sciences*.

Chase's letter describes "The Five Sexes" as "unique and refreshing" but critiques Fausto-Sterling for her use of the terms *true hermaphrodite*, *female pseudohermaphrodite*, and *male pseudohermaphrodite*. She maintains that such terms "reflect the Victorian belief that human sexual nature rests entirely in the gonads, a concept of gonadal determinism belied by the relative success of intersex medicine in sex reassignment" (1993, 3). As significant as the content of the letter is the fact that Chase identified herself as affiliated with ISNA, an organization she created in order to gain credibility. When her response was published, intersex people began contacting her to inquire about membership. Chase recalls: "Pretty soon . . . I'm getting all these letters from intersex people. They're coming into the P.O. box and I'm calling them and when I call them, they want to tell me their story, and they want to be on the phone for hours and hours and hours because they never imagined in their whole life that they'd ever meet anybody else. So I bought a headset . . .

[and ISNA became a reality]." Clearly, there was a need for community, both personal and organizational.

Just a few years after Chase fielded that outpouring of handwritten notes and telephone calls, the Internet created a new venue for community, connecting people who otherwise might never have encountered one another (see Still 2008; Preves 2005). The Internet was incredibly important in the early years of intersex advocacy, both for bringing people together to form a unique biosocial community and for self-education through the employment of an active biological citizenship. Sociologists Nikolas Rose and Carlos Novas (2005) define biosocial community through its functions: "to spread information about the condition; to campaign for rights and combat stigma; to support those affected by the illness; to develop a set of techniques for the everyday management of the condition; to seek alternative forms of treatment; and to demand their own say in the development and deployment of medical expertise" (448–49).

Without the Internet, intersex people might never have come together into what became a powerful biosocial community. Chase explains that ISNA was born "right at the beginning of when the Internet was beginning to be available." She also "remember[s] teaching people how to use e-mail and stuff." In "Making Media: An Intersex Perspective" (1997), she elaborates:

> The growth of the Internet has been a great boon for us. We have been able to leverage our computer skills into high visibility, making it very easy for intersexuals, parents, journalists, and professionals to find us. This visibility amplifies the impact each time major media [cover] us. Internet mailing lists have made it easy for us to bring our issues to professional communities outside the medical profession. Our Web site serves as an information clearinghouse not just to intersex people but [also] journalists looking for background research. It also promotes coverage already received, giving our struggle increased [cachet] among mainstream media. It is not an overstatement to say that without the Internet, it would have taken decades to get where we are now.[14]

Intersex people enacted biosocial activity by collectively forming their community through informational biocitizenship (developing specialized understandings about their bodies) and digital biocitizenship (connecting with similarly situated others through ISNA's e-mail list and website), among other forms of activism. According to Rose and Novas, this mode of engagement and community formation is indicative of an "active scientific citizenship, in which individuals are taking a dynamic role in enhancing their own scientific—and especially biomedical—literacy" (2005, 446).[15]

ISNA's membership grew exponentially and eventually included a mailing list of approximately 3,500 names and more than 1,000 donors (Preves 2005). With such numbers, ISNA became a major player in the U.S. intersex rights movement. Although ISNA was a virtual space wherein intersex people could connect with one another, it was not the only U.S. organization to play an important role in intersex advocacy, as AISSG-USA was also thriving.

Sherri Groveman Morris founded AISSG-USA in late 1995–early 1996. Its founding purpose was to reduce the stigma of intersexuality by connecting intersex people. Morris states: "Having found help for myself [through the U.K. AISSG group], I felt it important to ensure that other women with AIS [in the United States] have access to the same kind of information and support. But there was a second, and perhaps even more important, motivation for forming the group: to ensure that other adolescents not spend years feeling alone and afraid, burdened by a secret and unable to form healthy friendships and romantic attachments" (2006, 9). From early on, AISSG-USA was also concerned with improving intersex medical care. However, unlike ISNA, the main focus of which was advocacy, AISSG-USA had as its primary goal to offer emotional support to intersex women and their families. Still, although not all AISSG-USA members would view it as an activist organization, I approach it as such given that it serves as an educational organization that advocates for intersex people and their families by offering emotional support while simultaneously sharing medical knowledge with its members.

Since 2008, AISSG-USA has transformed itself in three key ways. First, in the summer of 2011, with the launch of its new website (www. aisdsd.org), the organization renamed itself AIS-DSD Support Group for Women and Families. This change made the group more inclusive by extending its scope to other intersex traits besides androgen insensitivity syndrome (AIS). (Because the majority of my data collection occurred while the organization was named AISSG-USA, I largely refer to it as such; I call it the AIS-DSD Support Group for Women and Families when discussing the period between the summer of 2011 and the 2013 annual meeting, after which point I refer to it as the AIS-DSD Support Group.)

Second, in the summer of 2013, and for the first time in its history, the organization opened its annual meeting to men with intersex traits; they not only were allowed to attend the meeting but even had "men's only" workshops and sessions. Organization membership remained limited to women with intersex traits and their parents, a policy that, as described below, was changed shortly thereafter. On the practical level, an "All Gender Restroom" was made available at the 2014 annual AIS-DSD Support Group meeting (see Figure 2.1).

Policy-wise, members agreed to hold a referendum vote, in late 2014, as part of the annual general election, on opening membership to anyone personally affected by intersex, regardless of gender identity or expression—that is, to men with intersex traits and to people with intersex traits who identify as gender queer or reject gender labels altogether. This referendum vote, which passed with overwhelming support, was a remarkable development, as the organization previously had a strict women-only policy, with the exception of male parents of children with intersex traits. AISSG-USA was formed as a support group for those with AIS, the majority of whom are assigned female at birth and identify as women throughout their lives. One reason for the initial reluctance to include intersex men may have been that these women, once they discovered they had XY chromosomes, might not have wanted any further association with intersex men, for fear of further undermining an already precarious sense of themselves as women. In 2012, however, Cheryl Chase, who now uses the name Bo Laurent, delivered the keynote address at the annual meeting and challenged the organization to include men born with intersex traits. That challenge became personal in 2013, when a teen member decided to gender transition. Without a bending of the existing policy, he would not have been able to attend the annual meeting. Many members considered it unethical and inappropriate to deny support to this young man and his family because he had gender transitioned, and this sentiment played a significant role in opening up the meeting.

Figure 2.1. A sign of change: a gender-inclusive restroom at the 2014 AIS-DSD Support Group annual meeting.

The third major change since 2008 has to do with the composition of the organization's board. The AIS-DSD Support Group board is made up of nine volunteer members and includes three officer positions: president, secretary, and treasurer. Parents of children with intersex traits were for years allowed to serve as general board members only. However, at the 2014 annual meeting, members called for a referendum to allow parents to hold officer positions in the organization. This change, which along with the other referendum on the ballot was passed with overwhelming support, is important to the organization's development, as it is evidence that intersex people and parents are working together in the organization in order to provide support and educational information for all of its members—not only those born with intersex traits.

AISSG-USA was only one of the groups that developed alongside ISNA. In 2003, Curtis Hinkle, a man with an intersex trait, started OII. Hinkle wanted to connect intersex activists around the world, as ISNA's membership was primarily U.S.-based, but he was also alarmed by the direction in which ISNA was moving. Not only was ISNA quickly shifting away from confrontational mobilization strategies such as public protests against the medical profession, but it was also welcoming medical experts into the organization in ways Hinkle believed were negatively influencing the direction of intersex advocacy. At the time of my field research, OII's mission was to create "a decentralised network established to give voice to intersex people both outside and inside the USA, those speaking languages other than just English, and people who do not fit the medicalised categories of disorder promoted by some other intersex groups. It is for people born with bodies which have atypical sex characteristics. OII resists all efforts to make intersex invisible, including genital mutilation, medicalisation and normalisation without consent and offers another face to intersex lives and experience by highlighting the richness and diversity of intersex identities and cultures."[16] Today, OII claims to be the largest activist organization within the intersex rights movement (a claim I have not been able to verify but have no reason to doubt).

Like AISSG-USA, OII has undergone substantial organizational changes over the years. First, the OII Network has spawned twelve global affiliates, all connected over the Internet: OII Australia, OII Austria–Vimö, OII Belgium–Genres Pluriels, OII Castellano–Intersexualidades, OII Chinese, OII Europe, OII Francophonie, OII Germany–IVIM, Intersex South Africa, OII Uganda–SPID, OII United Kingdom, and OII United States.[17] (During the bulk of data collection, OII was not as decentralized as it is today, thus I refer to it as OII except when discussing more recent events, in which case I reference the relevant OII affiliate.)

The second major organizational change has to do with OII's approach to collaboration. According to its website: "OII-USA promotes human rights for all intersex people, particularly the right to self-determination and bodily integrity. We strive to create a world where intersex people are viewed and treated equally by: providing information concerning actual life experiences of intersex individuals, and the goals of the global intersex advocacy community, *to all those working with intersex people*; supporting intersex individuals by providing information and contact with other intersex people from various groups and geographical regions; and assisting families and friends of intersex individuals in their role as allies, and in understanding intersex variations" (emphasis added).[18] This important shift, which we might view as evidence that OII is willing to work with others outside of the intersex community, including providers, happened sometime after I left the field in 2011. OII and its affiliates may be willing to offer information "to all those working with intersex people," but they will not defer to medical authority and its pathologizing DSD terminology: "All members of the Organisation Intersex International—OII—the largest intersex organization in the world, reject the label Disorders of Sex Development—or DSD—for the simple reason that we are not disordered, but different, and we refuse to accept medical language and views which pathologize us. The fact that some intersex individuals choose to use this term to describe themselves, just as some homosexuals view their homosexuality as a disease to be cured, does not discount the fact that the label is inaccurate and stigmatizing to the community as a whole."[19] Unlike the other organizations I studied, OII has always been and still is openly critical of the medical profession's treatment of intersex—a point I return to later in this chapter.

Accord Alliance is the final organization in my study, and it has the greatest historical connection with ISNA. By 2008, ISNA was widely accepted by intersex people and had become the largest intersex activist organization in the world. It was thus quite a shock when the organization closed its doors that summer. ISNA's closure and the birth of Accord Alliance are discussed in greater detail later in this chapter, but, in short, ISNA's leaders, especially founder Cheryl Chase, felt that the organization could no longer be a successful vehicle for change because of its history of confrontational mobilization strategies against the medical community. Accord Alliance emerged not long before ISNA ceased operations, and several former ISNA key players, including Chase, served on its inaugural advisory board. Its founding mission was simple: Its organizers sought to work in accordance with, rather than against, medical professionals in an alliance to promote change. The new organization's strategy was to educate medical professionals on their terms, a task that would have been difficult for ISNA to accomplish, given its history of protesting against and criticizing the medical profession.

Each of these intersex social movement organizations had its own strategy to address the shame and stigma associated with intersex traits. Each of those strategies in turn reflects a particular understanding of the gender structure. Before it closed its doors, ISNA enacted confrontational mobilization strategies and challenged simplistic ideas about gender, according to its belief that, as Chase (1997) put it at the time, "Gay men, lesbians, bisexuals, transgender people and intersexuals' oppressions stem from the same source: rigid cultural definitions of sex categories, whether in terms of behavior, identity, or anatomy."[20] Similarly, OII-USA sees intersex as a gender issue and is openly critical of the gender structure and of medical professionals who appear to enforce it in the ways they approach intersex traits.[21] OII continues to confront the medical profession publicly to promote social change. In contrast, Accord Alliance and the AISSG-USA/AIS-DSD Support Group do not see, or at least do not wish to publicly frame, intersex as a gender issue and thus on the surface do not challenge gender; instead, they have adopted an unobtrusive and collaborative approach to working with the medical community to promote change in intersex medical care.

Collective Confrontation

The early years of intersex activism focused on identity and involved intersex people's coming together to challenge the medical profession's approach to intersexuality.[22] I label this period, from the birth of the U.S. intersex rights movement to the turn of the twenty-first century, "collective confrontation." Through the collective efforts of ISNA activists and supporters, many of whom also belonged to other organizations such as AISSG-USA, the intersex rights movement pressured the medical profession to reconsider its medically unnecessary surgical and hormonal treatment of intersex traits. While feminist scholarship fueled the formation of the intersex rights movement, confrontational mobilization strategies were equally important to its progress, as they pushed medical professionals to acknowledge the voices of intersex people in the construction of their guidelines (e.g., Chase 1998a,b). ISNA activists routinely protested outside pediatric medical association meetings to raise awareness about the surgical modification of intersex genitalia (Karkazis 2008; Preves 2003). They also reclaimed the language of *hermaphrodite* by promoting its usage within their community and using it as a protest strategy. The original title of the ISNA newsletter was *Hermaphrodites with Attitude*,[23] and many participants wore t-shirts and held banners with the same slogan as they protested at meetings of medical professionals, gaining substantial media attention (see Figure 2.2).

David, a sixty-two-year-old early member of the movement, described these protest activities, revealing just how grassroots the movement was: "I remember [Cheryl Chase] saying to me once in the car, will you help me create the Intersex Society of North America, will you help me you know, be this movement? And . . . she would get people together to go to . . . pediatric conferences and protest outside them . . . and she had this huge banner that said 'Hermaphrodites with Attitude.'" Chase elaborated: "Any movement that can't get attention needs to do some attention-getting things, which we did." These "attention-getting things" often involved heated discussions and debates with medical professionals. Kimberly, a thirty-eight-year-old with an intersex trait, recounted one such confrontation at a Neonatal Nursing Conference, where a doctor was reporting on his study of women with congenital adrenal hyperplasia: "He had all this proof about how not disclosing, doing the surgery, and this stuff was beneficial. We had this fight, like a literal fight . . . it was a doctor, and I don't remember who he was. I'm kind of surprised it didn't come to blows, because I was ready to hit him. We just yelled at each other for like twenty minutes; it was terrible. I was completely unprofessional as was he."[24]

Intersex activists were angry, and rightfully so, because they felt violated and manipulated by medical professionals who had often kept their diagnoses from them when they were children. For example, forty-two-year-old Hannah didn't find out she had an intersex trait until she was in a college biology classroom:

> I was doing a buccal smear in college . . . when you swab the inside of the cheek . . . and it came up 46 XY. And I'm like *what*? So I had to do some investigation and went to the teacher and asked about it, and she was kind of like "I'm not sure" and I don't know if she knew and was like "I'm not saying anything" or what, but I basically had to go research what I had. She was just like "well, write down your results whatever they are," kind of indifferent about it . . . I knew I couldn't have children but I thought that I had had some kind of hysterectomy when I was 13 because I was told that my ovaries were precancerous and they had to be removed. So I was always worried that the cancer was going to come back or some-

Figure 2.2. Wearing a piece of intersex history. Here I am pictured with a now-historical t-shirt that was given to me by a leader in the intersex community.

thing. Now I know I was worrying for nothing. But I went to the library in school and I found some books and that's where I found the term "testicular feminization"[25] . . . and I was like, oh my God, that's what I have.

The efforts of early intersex activists eventually brought about change in this particular area, as the medical profession denounced hiding diagnoses from intersex people.

Although clinicians initially resisted intersex activists (e.g., Gearhart 1996; Meyer-Bahlburg 1996), ultimately they began to pay attention. By 2000, Chase was invited to speak at the then-named Lawson Wilkins Pediatric Endocrine Society (LWPES),[26] a group she had earlier protested. This successful activist encroachment onto medical turf was highly unusual, for two overlapping reasons. It marked the first time organizers of a major U.S. medical conference solicited an activist's perspective on intersex (Karkazis 2008). It was also "the first time that the society's annual symposium was devoted to intersexuality" (Karkazis 2008, 257). Dr. D., a pediatric endocrinologist, concisely summarizes the medical profession's eventual understanding of early intersex activism: "So the original folks who self-aggregated *hated* the medical community, or were very angry with them, not inappropriately, for the way that they had been treated."

Also by 2000, after almost a decade of protests by adult intersex activists, the American Academy of Pediatrics issued a statement titled "Evaluation of the Newborn with Developmental Anomalies of the External Genitalia" (Committee 2000), which can be seen as definitive proof that 1990s intersex activism successfully brought about social change. Not all the suggestions presented in the medical statement were implemented, nor were the recommendations flawless. Although the guidelines state that intersex infants "should be referred to as 'your baby' or 'your child'—not 'it,' 'he,' or 'she'" (Committee 2000, 138), which was undoubtedly progress, they still included surgery as a viable approach to dealing with intersex traits. Doctors were advised to inform parents that their baby's "abnormal appearance can be corrected and the child raised as a boy or a girl as appropriate" (Committee 2000, 138). Among the factors to be considered when determining which "gender assignment" should be recommended for a given intersex child were "fertility potential" and "capacity for normal sexual function" (Committee 2000, 141), concepts and language that revealed still-pervasive biases.

Many medical professionals adhere to cissexist and heteronormative ideologies that are especially problematic in the case of the medical management of intersexuality, as they are imposed on intersexuals and their bodies during gender assignment. Cissexism is the belief that gender is authentic only when it neatly aligns with sex (Serano 2007), and heteronormativity is "the suite of cultural, legal, and institutional practices that maintain normative assumptions that there are two and only two genders, that gender reflects biological sex, and that only sexual attraction between these 'opposite' genders is natural or acceptable" (Schilt and Westbrook 2009, 441; see also Kitzinger 2005). If an intersex person's gender assignment is based on capacity for "normal [hetero] sexual function" and "fertility potential," for instance, those born with congenital adrenal hyperplasia (CAH), who are capable of pregnancy despite having a large clitoris that resembles a small penis, will be defined as female and regarded as in need of surgery, regardless of the fact that they might desire to penetrate using their phallus.[27] In other words, the medical professionals who wrote the 2000 statement evinced a heteronormative ideology by failing to recognize any possibilities beyond gender- and sexual-normative bodies in their gender and medical intervention recommendations.

Although the statement wasn't perfect, it nevertheless stands as evidence that the medical profession was beginning to listen to intersex activists who were gaining increasing credibility. Chase explained:

> Working in concert with more established groups, we generated credibility by association. The Gay and Lesbian Medical Association carried some of our releases in their newsletter and invited us to present a panel at their annual meeting in 1996. The Association of Gay and Lesbian Psychiatrists helped us to set up a case presentation and open house at the APA's annual meeting in spring 1997. Perhaps even more significant than the education of a few dozen physicians that was accomplished at each of these events was the fact that media took us more seriously because we had gained a medical audience. These are also news events, particularly for professional publications. American Education Gender Information Society (AEGIS) sent out an issue of *Hermaphrodites with Attitude* in their August 1995 mailing to 600 professionals. GenderPAC and AEGIS carried us on Internet [listservs]. GLAAD began to feature us in its publications, which amplifies our work in two ways. One is to more reliably deliver our releases to a large number of journalists. Journalists skim GLAAD's lists; they are more likely to toss a release from an unknown organization. *GLAADAlert* also provides feedback to the media for coverage of intersex. The producers of the intersex segment on *Dateline* told us that the piece generated a great deal of favorable viewer response; much of that can probably be attributed to GLAAD.[28]

This media attention allowed activists to frame intersex as a social rather than a medical problem. However, it was also a direct challenge to medical authority and jurisdiction over the intersex body.

The Birth of Disorder of Sex Development (DSD)

In August 2005, Alice Dreger, Cheryl Chase, Aron Sousa, Philip Gruppuso, and Joel Frader published "Changing the Nomenclature/Taxonomy for Intersex: A Scientific and Clinical Rationale" in the *Journal of Pediatric Endocrinology & Metabolism* (Dreger et al. 2005). The authors, all known ISNA allies and supporters of Chase, concluded: "[W]e suggest the language of 'hermaphroditism' and 'pseudohermaphroditism' be abandoned. One possible alternative to the procrustean 5-sex approach is to use instead specific etiology-based diagnoses (such as AIS, 5α-reductase deficiency, etc.) and the umbrella term 'disorders of sexual differentiation.' Such an approach would have the salutary effects of improving patient and physician understanding and reducing the biases that are inherent in the use of the current language of 'hermaphroditism'" (Dreger et al. 2005, 733).[29] A few months later, in October 2005, fifty international experts on intersex, with different types of expertise, gathered and developed the "Consensus Statement on Management of Intersex Disorders" (Houk et al. 2006; Hughes et al. 2006; Lee et al. 2006). Chase was present at the meeting and, as we will see, played an instrumental role in the consensus statement, which was simultaneously published in multiple venues (Houk et al. 2006; Hughes et al. 2006; Lee et al. 2006). The conference was jointly organized by Dr. Peter Lee of the Lawson Wilkins Pediatric Endocrine Society and Dr. Ieuan Hughes of the European Society for Paediatric Endocrinology. However, this was not your traditional academic conference where researchers come together to share findings from their latest projects. Instead, conference attendees were organized into six groups, each of which had between four and six members and two coordinators. Each group received in advance six prompts that they would address at the conference. The focus of the groups was medical, and their topics ranged from "Recent Molecular Genetic Impact of Human Sexual Development" to "Surgical Management of Intersex." DSD nomenclature did not appear on the conference agenda[30] and did not come up until Dr. Eric Vilain, a geneticist at the University of California, Los Angeles, raised the possibility of new terminology, disorders of sex differentiation (DSD), in what has been labeled a "rushed final plenary session."[31]

But Vilain did not act alone. According to Chase, "[I] engineer[ed] the entire thing . . . getting the language changed into [DSD terminology] by working through allies. There were some progressive people in there, and there were some powerful people who were on my side, and I talked to them about what ISNA thought we would like to have happen." Chase explained that she needed Vilain and other allies to introduce the new nomenclature at the consensus meeting because medical professionals weren't willing to entertain suggestions from an intersex person with a long history of confrontational intersex activism:

They wouldn't listen to me. I was put in a group where I couldn't do any harm. I talked to people in the other groups and sort of back-channeled, and said, "This is what I want to get done. Can you get it done in your group?" . . . I had draft copies of the handbooks with me. I said, "I want a language that is really appealing to doctors, that fits somehow into the way they think about things, so it'll be easy to adopt." I said, "I want it not to have the word 'pseudo.' I want it not to have any indication of sex or gender. I want it not to have the word 'hermaphrodite' or 'true.'" At that point, in [ISNA's] handbook[s] it said "disorders of sex differentiation." And then they went off and they decided there were some technical reasons why they preferred "development" to "differentiation," and they got it adopted. Then we changed [ISNA's] handbooks to match.

At that time, ISNA had two handbooks meant to educate readers about intersex—one for clinicians (Consortium 2006a) and one for parents (Consortium 2006b)—in the final stages of publication. They replaced intersex language with the new terminology, a move that angered many of those involved in the organization. Several people who had shared their personal stories in the handbooks, including David Cameron, Peter Trinkl, and Esther Morris Leidolf, demanded a disclaimer stating that they did not agree with the disorder language. Despite these protests, the wording of the handbooks was a noticeable move away from the collective confrontation of the medical community that dominated 1990s intersex activism.

In May 2006, in light of the Chicago meeting, the "Consensus Statement on Management of Intersex Disorders" was published (Houk et al. 2006; Hughes et al. 2006; Lee et al. 2006). The revision of the 2000 statement was necessary because of "progress in diagnosis, surgical techniques, understanding psychosocial issues, and recognizing and accepting the place of patient advocacy" (Lee et al. 2006, 488). The new recommendations included avoiding unnecessary surgical intervention: "Because orgasmic function and erectile sensation may be disturbed by clitoral surgery, the surgical procedure should be anatomically based to preserve erectile function and the innervation of the clitoris. Emphasis is on functional outcome rather than a strictly cosmetic appearance. It is generally felt that surgery that is performed for cosmetic reasons in the first year of life relieves parental distress and improves attachment between the child and the parents; the systematic evidence for this belief is lacking" (Lee et al. 2006, 491). Although this statement about irreversible surgical interventions was promising, there was still no guarantee that medical professionals would follow its recommendations.

The consensus statement also recommended a nomenclature shift: "Terms such as 'intersex,' 'pseudohermaphroditism,' 'hermaphroditism,' 'sex reversal,' and gender-based diagnostic labels are particularly controversial. These terms are perceived as potentially pejorative by patients and can be confusing to practitioners and parents alike. We propose the term 'disorders of sex development' (DSD), as defined by congenital conditions in which development of chromosomal, gonadal, or anatomic sex is atypical" (Lee et al. 2006, 488). I discuss the medical profession's openness to this recommendation in chapter 3. Here, what matters is that this new DSD language forever changed the intersex community. Because the powerful medical community formally introduced it in their consensus statement, intersex people and our families[32] were forced to engage with it.

In 2010, only four years after this consensus statement was published, Dr. Ieuan Hughes, a co-organizer of the Chicago meeting, noted, "The DSD nomenclature and its spin-offs have arrived at the high altar of medical practice. The revolution may have been quiet, but it certainly has been effective and achieved with the minimum morbidity" (Hughes 2010a, 161; see also Hughes 2010b). Dr. Ian Aaronson and Dr. Alistair Aaronson similarly acknowledged, "The term disorders of sex development (DSD) has achieved widespread acceptance as replacement for the term intersex" (2010, 443). Other accounts support this claim[33]— which was never Chase's goal, as she hoped it would replace only terminology that used or incorporated forms of the word *hermaphrodite*. While Hughes frames this renaming as an all-encompassing and victimless victory, we must not forget that he speaks from a medical perspective, rather than from the perspective of those personally affected by intersex traits. This shift in diagnostic language had major implications throughout the global intersex community, marking what I refer to as the shift from collective confrontation against the medical profession to contested collaboration with the medical community.[34]

Contested Collaboration

Intersex people have always been critical of the medical management of intersexuality, particularly irreversible surgical intervention and the lack of honest and complete diagnosis disclosure. ISNA was formed on the basis of this critique, and its mission was always very clear in this regard. As its website states, the organization was "devoted to systemic change to end shame, secrecy, and unwanted genital surgeries for people born with an anatomy that someone decided is not standard for male or female."[35] Expanding on this base, the website explains that ISNA "learned from listening to individuals and families dealing with intersex" that "parents' distress must not be treated by surgery on the child; professional mental health care is essential; honest, complete disclosure is good medicine; all children should be assigned as boy or girl, without early surgery."[36] Yet in the summer of 2008, the world's largest intersex social movement organization announced that it was closing its doors, releasing the following statement to explain its decision: "This is ISNA's dilemma: we finally have consensus on improvements to care for which we have advocated for so long, but we lack a consistent way to implement, monitor, and evaluate them. At present, the new standard of care exists as little more than ideals on paper, thus falling short of its aim to improve the lives of people with [disorders of sex development] and their families."[37] Chase and other advocates who supported shutting down ISNA believed that changing actual medical practices would occur more easily if they worked with, rather than against, medical professionals. Many argued that ISNA had been tainted by its confrontational past. Chase explained:

> As we begin to have an audience with parents and doctors, we need to be able to talk with them [and] not from a radical stance. [However, they] wouldn't allow us to become more moderate . . . they just wouldn't believe it. There were people outside the healthcare system who just wanted to keep picketing doctors. There were some people who picketed healthcare conferences after we were already inside speaking at them, and that doesn't help. We came to realize that the reputation that we had had been an asset in early days but it had become a negative asset. It prevented us from talking to other people that we cared to talk to, and that's why we decided to change.

Jane, a fifty-four-year-old individual with an intersex trait, had similar thoughts: "Although to demonize members of the medical profession might be part of the history of the intersex rights movement, it certainly did something to raise awareness and start people talking about it, but it does absolutely nothing, quite the contrary, as far as changing the standard of care [goes] and it certainly doesn't open up a dialogue when you're calling people names and threatening them with personal harm and, um, screaming and shouting." However, this decision to close ISNA was made without input from all of its members, and it came as a horrific surprise to many. Millarca, a forty-six-year-old longtime ISNA activist, wasn't aware that ISNA had closed until she found out second-hand: "I called Cheryl Chase and said, 'What's going on?' She said, 'We're closing ISNA.' I said, 'Are you fucking kidding me?' She said, 'Yeah, we're closing ISNA, we're opening a more conservative organization that's going to have the medical profession accept intersex and DSD more. And we're changing the name from *intersex* to *DSD*.' I was like, 'Really?'" Other intersex activists intended to contact Chase and voice their disappointment but hesitated, realizing the damage had already been done. Mercurio, a forty-one-year-old intersex activist, described her thinking at the time: "And I actually started a letter to Chase, and I never sent it, basically personally asking her to drop the term 'DSD,' to just change it. I mean, it's an online document. I don't know, she probably needs the approval of all the doctors on the consensus committee, so actually that would probably be pretty difficult, unfortunately, at this point, or at least to change it to 'differences.'"

In 2007, roughly a year before ISNA shut down, a few influential members, including Cheryl Chase and Arlene Baratz (a parent in the community who is a doctor, though not in intersex care), "sponsored and convened a national group of health care and advocacy professionals to establish a nonprofit organization charged with making sure the new ideas about appropriate care are known and implemented across the country."[38] By March 2008, Accord Alliance had been formed, with the goal of working with the medical profession to promote change. Although activist Kimberly wasn't involved in the formation of Accord Alliance, she understood and supported the collaborative model they were striving for:

I needed to be able to get them involved and on my side and that just being confrontational with them and having an "us and them" attitude wasn't doing me any good. So it was a real turning point for me of recognizing a bit more on how this has to be done, and how to have the medical community . . . they need to be allies. I guess I came from a place originally where they were the enemy. The first year or two, I remember the most I could kind of come up with was well, maybe they're not enemy *enemy*, but because of their incompetence, how many people have suffered? . . . They might have gone in with the best of intentions, but their incompetence was screwing up people's lives, and that's what it was. [From there, I went] to recognizing these are people who are trying to do the best thing, they need more information, but we need to speak their language to get that information across. So that was an evolution for me personally and I think it was an evolution in the community as well.

Whereas ISNA engaged in confrontational mobilization strategies to promote change in intersex medical care, Accord Alliance defines its mission as "promot[ing] comprehensive and integrated approaches to care that enhance the health and well-being of people and families affected by DSD" and seeks to accomplish this "by fostering collaboration among all stakeholders," including medical professionals.[39] While Accord Alliance shares ISNA's goal of enhancing the well-being of those born with intersex traits and their families, its strategy clearly differs. ISNA opened the door, as Jane put it, "So [now] the door is open and the discussion starts. So as a pragmatist, do you understand how powerful that is?" Many medical professionals shared Jane's sentiment and openly embraced the shift in mobilization strategies. Dr. D. summarized this change: "Over time, there were individuals who didn't hate the medical community so much and wanted to make things better for their children, or their peers, or their siblings, and those who followed after them had said, 'We cannot just be angry, we have to do something.'"

The unobtrusive mobilization strategies currently employed by both Accord Alliance and the AIS-DSD Support Group are based on "discursive politics" and "occupy and indoctrinate" tactics. Discursive politics "includes talking as well as the production of tangible or symbolic materials that are aimed at political change . . . the telling of stories about, for example, rape victims, crimes, and justice" (Schmitt and Martin 1999,

369). A group (or individual) that occupies and indoctrinates works "*from* the inside *to* the inside, in order to imbue the mainstream with its political understandings" of the social issue at hand (Schmitt and Martin 1999, 369). Working within the gender structure, Accord Alliance and the AIS-DSD Support Group do not seek to promote change and awareness about intersexuality by publicly challenging binary understandings about bodies. Instead, they disseminate stories and experiences of trauma. Accord Alliance focuses on educating the medical community, while the AIS-DSD Support Group focuses on offering resources and support to intersex people and our families. In doing so, as noted previously, the AIS-DSD Support Group has recently—and controversially—been working with the medical community to bring about change by holding a continuing medical education (CME) conference in conjunction with their annual meeting and inviting medical professionals to speak at the annual meeting about everything from hormone replacement therapy to the latest medical research on intersex traits.

Not all intersex people in the community embrace these collaborative mobilization strategies. Millarca, the intersex activist quoted earlier, emphatically stated, "I don't have a relationship with Accord Alliance. I know the people that are affiliated with Accord Alliance. I see them at conferences and I'm polite and I say, 'Hi.' I don't believe in an organization that is trying to help kids by having medical professionals on the board whose purpose is against what our focus was in the beginning." Mercurio, the activist who did not support the move to DSD, expressed similar distaste: "Working with the medical community has become working for the medical community because there's been no specific gains. And now, the pathologizing terminology, which originated in the medical community because some practitioners were using 'disorder' already, has been stamped upon us so it's like, what have we got out of this?" Chris, a fifty-seven-year-old intersex activist, was even more critical of medical professionals:

> I basically see the medical community as it applies to intersex like a cult, a religious cult. It's like a high priest in white coats, and they're performing rituals to initiate the unclean and the unwashed into this mythical male/female. And they know this is bullshit, some of them. But some of them are still blinded. Doctors are not assigning a sex. This is all decep-

tive. It's a hoax. They're assigning a gender identity to children because if they were assigning a sex, in many cases they wouldn't be able to decide our sex by the current standards. What they're actually putting on birth certificates is not a sex, it's a gender identification and then they're going to start transforming our bodies by what gender identity they think we should develop.

Chris explained his understanding of the mobilization shift, specifically his criticism of ISNA and its alliance with medical professionals:

ISNA, I felt it started out as somewhat of a support group and it gradually morphed into an organization that seemed basically just interested in talking to doctors, not providing support, and they seemed to be wanting to talk to doctors because they wanted to change the current treatment. And I felt like what happened is that they became too devoted to the donors, many of [whom] were probably doctors, and they were in such a dialogue that they had lost any close relationship with the people they were supposedly advocating in favor of, which was the intersex people. . . . And see, the problem is, once a marginalized group starts letting people be in an organization on a board with people that have elevated social status, it's going to be very hard to not have a lot of the members feel subservient, and they can slowly take over without your being aware of it. . . . If you start accepting large amounts of money from medical people, well, that could be a real conflict of interest. . . . I felt like ISNA became too doctor-focused and too medically focused to where they lost contact with the grassroots. I feel like that's what happened. They were talking to doctors, but they weren't talking to any of the rest of us. And I think the doctors are part of the problem.

After Accord Alliance replaced ISNA, confronting medical professionals would seem to have become a thing of the past for the intersex rights movement. A parallel transition happened when AISSG-USA changed its name; AISSG-USA never engaged in confrontational mobilization strategies, and the AIS-DSD Support Group has even more reason not to, because the organization has been and continues to be incredibly successful. Its membership is larger than ever and it is well known throughout the global intersex community. The exception to this rule is OII. OII remains skeptical of the collaborative approach favored by Accord Alliance and the AIS-DSD Support Group and does not hesitate to publicly criticize medical professionals. OII-USA's website states: "Surgical 'normalization' of intersex bodies is an attempt, like eugenics, to remove differences which some people have decided are undesirable, and it often creates problems which were previously non-existent. To view intersex variations as 'medical conditions' which can be cured justifies the barbaric medical practices we are often subjected to, such as genital surgeries and/or hormones which may be contrary to our own core identity, and psychological treatments for not wishing to comply."[40]

Of the intersex social movement organizations in this study, OII is the only one that today openly, directly, and confrontationally suggests that the problems associated with intersex traits have to do with narrow views about gender. To use Raewyn Connell's terms (1987), OII is actively disrupting the medical profession's "gender regime"—that is, its perpetuation of heteronormative ideologies based on the idea that women and men are innately different. According to the OII-USA website, "To many intersex people, gender is the main issue. To erase the importance of gender to the individual intersex person is to reduce that person to only the physical aspects of their body, neglecting the more important part of the equation, their own perception of that body and themselves, as opposed to how others perceive them."[41] By framing intersex as a gender issue that medical professionals tend to control through "surgical normalization," OII challenges medical authority in much the same way ISNA did when it began. The medical profession, by and large, is a powerful group that enforces gendered ideologies about bodies and identities. In the next chapter, I will show that many medical professionals hold essentialist and binary ideologies of sex, gender, and sexuality, which they rely upon when making recommendations and performing procedures on intersex people. These prominent medical ideologies conflict directly with OII's views, which leaves these two groups talking past, rather than with, each other.

Conclusion

The development and progress of the intersex rights movement and its social movement organizations manifests the reflexive interconnections between academic knowledge, technology, and human behavior (see Giddens 1990, Melucci 1989). Academics not only study society but shape it in ways that influence all of our lives; technology facilitates social change but can also serve as a vehicle for co-optation; organizations, like individuals, evolve, often with unforeseen consequences. Feminist scholarship and the advent of the Internet paved the way for the formation of ISNA in 1993. The resulting U.S. intersex rights movement sought to end the shame, secrecy, and stigma associated with intersex traits and allowed intersex people to learn about their diagnoses, connect with one another, and collectively confront the medical profession. Their successes, ironically, led to a transformation of intersex advocacy and a general move toward collaborating with medical professionals to promote change. One significant example of this was the 2006 introduction of DSD terminology, which, again ironically, has diminished some of the influence and progress achieved by 1990s intersex activists.[42]

Both the 2000 and the 2006 medical statements were indisputably evidence of progress: No matter how few or problematic their recommendations, and whether or not they were fully implemented, they represented some degree of positive change for intersex people. Further evidence of the success of early intersex activism can be found in women's and gender studies curricula and textbooks, almost all of which now discuss intersex. Since the 2006 consensus statement, which formally introduced DSD terminology, intersex advocacy has largely transformed from confrontational to collaborative mobilization strategies. While OII continues to challenge medical professionals at the institutional level of gender structure, Accord Alliance and the AIS-DSD Support Group work with medical professionals to promote change. There is no clear indication that one strategy is positioned to be more successful than the other. We cannot foresee the future, and it might just be the case that we need both strategies—those who push societal boundaries by challenging the gender structure and those who work within the gender structure to gain sympathy and public awareness about intersex traits— for real change to occur.

As for the new DSD terminology, I remain unconvinced that it was the right way to go. Although none of us, including Chase and her allies, who initiated it, could have predicted that DSD terminology would replace intersex language in ways that have affected our community so profoundly, it seems clear by now that medical professionals have strategically used the terminology as a vehicle to reclaim their jurisdiction over intersex, just as intersex activism was seemingly beginning to successfully challenge that authority. In the next chapter, I explore how this happened and look at how DSD terminology has forever changed our intersex community.

3

Medical Jurisdiction and the Intersex Body

It was a beautiful fall day and I was in a major U.S. pediatric medical center, practically running down a hospital stairwell. I was following Dr. I., a world-renowned expert on intersex, who was on her way to her DSD team's meeting, which was already in session. Dr. I. had agreed to an interview months earlier, but on the scheduled morning she had a family emergency that left her running late. Understanding that I had traveled across the country to interview her, Dr. I. was determined to help me in any way she could, so she graciously invited me to accompany her to the meeting. Though Dr. I.'s interview was one of the shortest I conducted during this entire project, she did her best to answer as many questions as possible, addressing several as we flew across the hospital.

It had been four years since the publication of the "Consensus Statement on Management of Intersex Disorders" (Houk et al. 2006; Hughes et al. 2006; Lee et al. 2006), and DSD terminology had replaced intersex language, at least in the medical profession.[1] When we entered the room, I saw a dozen or so medical professionals in white lab coats and hospital scrubs. Most were in-house DSD experts, and the rest appeared to be medical residents. They had convened to discuss the medical management of their DSD patients. Dr. I. told the team that she had invited me because I was "doing really great work" and then asked if I would say a few words about my research. I did so, a bit nervously, as I hadn't been prepared for such an opportunity. One reason for my nervousness was that I didn't know whether I should disclose that I had been born with an intersex trait. I feared that once these medical experts knew my personal connection to the topic, they would no longer be willing to be interviewed, but I was wrong. In the heat of the moment, I decided to disclose, and the group still received me warmly. One medical professional, who was not yet scheduled to meet with me, even asked for my contact information so that he could schedule an interview while I was still in town. My experience with Dr. I. and her DSD team was not unusual. I found that most[2] medical professionals I invited to participate in my study were not only supportive and helpful throughout data collection but also determined to provide the best possible medical care in their treatment of intersex.

In chapter 2, I described how, beginning in the late 1970s (see Kessler and McKenna 1978), feminist scholars critiqued medical interventions on intersex bodies by drawing attention to the social construction of sex and gender and, perhaps most important, the lack of complete and honest diagnosis disclosure, beneficence, and bodily autonomy afforded to individuals with intersex traits (see also Fausto-Sterling 2000a,b, 1993; Dreger 1998a,b,c; Kessler 1998, 1990). These critiques helped kickstart the intersex rights movement, bestowing scholarly legitimacy on intersex activism.[3] With feminist scholarship and intersex activism validating each other, the medical profession came under fire for its treatment of intersex traits. The flames of this fire were only further fanned in the late 1990s, when psychologist John Money, an influential authority on intersex medical care throughout the second half of the twentieth century, was exposed for using falsified data to uphold his gender theory, which in turn was harmfully used to justify surgical interventions on intersex bodies.

In this chapter, I argue that the medical profession holds a unique position at the institutional level of gender structure, where it can either perpetuate or challenge traditional gender ideologies and, in so doing, use its authority either to harm or to help intersex people. I construct this argument on two foundation blocks: the literature on the sociology of diagnosis, in particular the relationship between naming and treating medical traits (see, e.g., Jutel 2011, 2009); and current knowledge about the medical management of intersexuality. The medical community did not always have a framework for defining abnormalities. The ancient Greeks practiced medicine without diagnostic names, simply describing diseases (Veith 1969). This practice continued until the eighteenth century, when medical professionals adopted a botanical model of linguistic classification that identified and labeled a wide range of medical traits (Fischer-Homberger 1970). Although some medical professionals resisted the naming of diseases, the "classificatory project" prevailed as "medicine shift[ed] its focus from individual symptoms to groups and patterns of symptoms that doctors could reliably recognise" (Jutel 2009, 280–81; see also Jutel 2011). This new approach centered on classifying conditions under defined diagnoses. Of course, defining conditions depends upon context, including time, place, and available technology. For example, twentieth-century technological advances led to the discovery that males were allegedly distinguishable from females by their sex chromosomes: XY for males and XX for females (Moore and Barr 1955; Moore, Graham, and Barr 1953; Barr and Bertram 1949). Chromosomal testing introduced a new, albeit ultimately problematic, marker of sex and in turn a new way to identify intersex traits.[4]

Sociologist Phil Brown (1990) argues that the diagnostic process involves two components: diagnostic technique and diagnostic work. Diagnostic technique "involves formalization of classification, including the specific tasks, techniques, interviews, and chart recording necessary to make the formalized classification," whereas diagnostic work "consists of the process by which clinicians concretely proceed with their evaluation and therapeutic tasks" (Brown 1990, 395). This distinction helps us better understand intersexuality, as the discovery of sex chromosomes became a new formal way to legitimize sex classification. Although providers were performing the diagnostic work of surgical intervention on intersex bodies before sex chromosomes were discovered (Mak 2012; Reis 2009; Dreger 1998a), sex chromosomal testing offered a new way of classifying intersex traits that further justified existing surgical procedures.

These insights from the sociology of diagnosis are useful in understanding the contemporary medical management of intersex traits. They enhance our understanding of the institutional level of gender structure in ways that enable us to see: (1) how diagnosing intersex is connected to the belief that sex, gender, and sexuality are clear and correlated binaries; (2) how medical professionals' definitions of intersex sometimes have unintended negative consequences, notably the justification of medically unnecessary interventions on healthy bodies; and (3) how the renaming of *intersex* as a *disorder of sex development* allowed medical professionals to reclaim their authority and jurisdiction over intersex, which had come under fire from intersex activism, feminist critiques of intersex medicalization, and Money's exposure.

The Rise of John Money

Intersex people had been subjected to medical interventions since the nineteenth century (see Mak 2012; Reis 2009; Dreger 1998a), if not earlier (Warren 2014). However, the twentieth century was noticeably different because of John Money, a psychologist at The Johns Hopkins Hospital.[5] Money's work influenced gender scholars across disciplines, but his most important impact was on the medical treatment of intersex traits. Money and his colleagues initially published their work on gender socialization in the *Archives of Neurology and Psychiatry* in the late 1950s (Money et al. 1957). In 1972, Money and one of his students, Anke A. Ehrhardt, expanded upon these ideas in *Man & Woman, Boy & Girl: Differentiation and Dimorphism of Gender Identity from Conception to Maturity*. Although Money and Ehrhardt argued that the human brain was at least somewhat influenced by hormonal exposure during gestation (see, e.g., their chapters "Fetal Hormones and the Brain: Human Clinical Syndromes" and "Gender Dimorphism in Assignment"), they maintained that childhood gender socialization was the most significant factor in explaining an adult's gender identity.[6] They suggested that, regardless of their genitalia, babies could live happy and healthy lives if they were strictly gender socialized (Money et al. 1957; see also Money and Ehrhardt 1972). They supported their argument with their clinical data, mostly on intersex people, and concluded, "So much of gender-identity differentiation remains to take place postnatally, that prenatally determined traits or dispositions can be incorporated into the postnatally differentiated schema, whether it be masculine or feminine" (Money and Ehrhardt 1972, 103).

Gender scholars across disciplines used Money and Ehrhardt's empirical support for their gender socialization thesis to refute claims that gendered behaviors were biologically determined. Although gender scholars in nonclinical social sciences used different terminology, notably "sex role theory," their ideas about gender were in line with Money's theory. Sociologist Barbara Risman concisely summarizes sex role theory: "Sex role theory suggests that early childhood socialization is an influential determinant of later behavior, and research [under this theoretical framework] has focused on how societies create feminine women and masculine men" (1998, 14). Money's research, offering empirical evidence from a clinical setting, served as "hard" evidence for the sex role thesis. This is not to say, however, that his work met only praise. In *Myths of Gender*, Fausto-Sterling (1985) critiqued him and his colleagues for routinely basing their analyses on hegemonic gender and sexuality stereotypes.[7] Sex role theory, itself, was also eventually critiqued, which paved the way for gender scholars to conceptualize gender as a stratification system rather than as roles people are socialized into (see, for example, Martin 2004; Risman 2004, 1998).

Money's research also had implications beyond the academy, most notably in the medical management of intersex. Doctors confronted with intersex traits, especially outwardly ambiguous genitalia, could find solace in Money's theory. They could assign a gender and then surgically shape the genitals so that the assigned gender was in line with the surgically constructed sex, secure in Money's assurance that if the baby was socialized according to the assigned gender, everything would be fine.[8]

The medical profession was particularly fixated on Money's claims and research, as opposed to the sex role theory of nonclinical gender scholars, because they aligned most closely with the positivist paradigm that dominates the medical profession. Still, the fact that medical professionals were even engaging with social constructionist claims was quite remarkable. In a sense, the presence of intersexuality demanded such engagement. Surgically modifying genitalia for medically unnecessary reasons was possible only because doctors and parents believed that intersex infants could live normal lives if they were socialized in accordance with their surgically constructed sexes. This assumption necessitated a rather selective medical belief in gender socialization, but nevertheless it set the stage for interdisciplinary collaboration across the social and medical sciences.

In 1966, working with other medical professionals, Money established the Gender Identity Clinic at Johns Hopkins. Although the Clinic is now defunct, for decades it remained the state-of-the-art medical center for the treatment of intersexuality and transsexuality, and Money and his team received referrals from doctors across the country (Distinguished 1986). In 1985, the American Psychological Association awarded him its Distinguished Scientific Award for the Applications of Psychology. The April 1986 issue of *American Psychologist* summed up his "unparalleled contributions to theoretical analysis and clinical treatment in human sexuality":

> He originated the seminal concepts of gender identity and gender role which form a cornerstone in all modern theories of sexuality. His extensive contributions to theory and research are characterized by a biosocial perspective which combines animal experiments with ethnological reports, and by a longitudinal approach extending from prenatal sex determination to gender role changes during old age. His clinical contributions include effective hormonal treatments of male sex offenders, hypogonadal impotence, and of virilism in congenital adrenal hyperplasia. He has excelled in psychological management of families having children with symptoms of intersexuality. (Distinguished 1986, 354)

In the face of such a reputation, medical professionals could not ignore Money's gender dimorphic behavior theory. Even his harshest professional critic, Dr. Milton Diamond, a prominent and powerful professor of anatomy, biochemistry, and physiology at the University of Hawaii, had to at least engage with Money's claims (Diamond 1979, 1978, 1976, 1968, 1965; see also Gadpaille 1980; Imperato-McGinley et al. 1979; Zuger 1975, 1970). In 1982, for instance, Diamond strongly stated his objections to Money's research:

> My own research and clinical experience in dealing with homosexuals, transsexuals, and other individuals with various sexual and gender identities while on the Board of Directors of the Hawai'i Sexual Identity Center and as a medical school faculty member have led me to suspect otherwise [about Money's theory], and I have several times challenged the correctness of the theory and the validity of the recommended practice as well as [its] predicted [clinical] success . . . . Others, too, have questioned the theory and practice. (182)

Although Diamond and some of Money's other medical peers remained critical of his theories, he would routinely be cited as the premier proponent of the gender socialization thesis, perhaps not surprisingly, given that he and his colleagues published "approximately 300 scientific papers, 75 scholarly reviews and textbook chapters, and 11 books" (Distinguished 1986, 355).

The Fall of John Money

In the late 1990s, Money and his team faced their harshest critic, a former patient by the name of David Reimer. In *Man & Woman, Boy & Girl*, Money and Ehrhardt relied on an unprecedented case study of identical twins to support their gender socialization argument. Unlike the majority of patients treated by Money and his team, the twin boys highlighted in the case study were biologically "normal" males, born on August 22, 1965. Their parents did not consider circumcision until months later, when their mother noticed they were having difficulty urinating because of a treatable foreskin complication known as phimosis. On April 27, 1966, the twin boys were scheduled for circumcision. However, a surgical mishap left one of the boys, who would eventually be known as David Reimer, with a penis severely burned beyond repair. Worried that this tragedy would prevent David from living a normal life, his parents contacted Money for medical advice. Money assured them that he had a solution. He and his team would surgically remove what was left of David's penis and construct a vagina in its place. As long as the parents socialized their sex-reassigned child as a girl, their problem would be solved:

> In particular, [the parents] were given confidence that their child can be expected to differentiate a female gender identity, in agreement with her sex of rearing. They were broadly informed about the future medical program for their child and how to integrate it with her sex education as she [grew] older. They were guided in how to give the child information about herself to the extent that the need arises in the future; and they were helped with what to explain to friends and relatives, including their other child. Eventually, they would inform their daughter that she would become a mother by adoption, one day, when she married and wanted to have a family. (Money and Ehrhardt 1972, 119)

The baby was surgically modified, and, with Money's encouragement, the parents engaged in extreme gender socialization in an effort to reinforce the sex reassignment. Money published extensively on what became known in academic literature as the John/Joan case,[9] which was used for decades as sound empirical evidence to support the gender socialization hypothesis.

In 1997, Diamond's longstanding critique of Money gained traction,[10] and the credibility of the John/Joan case, and ultimately of Money himself, was challenged throughout the media including in *Rolling Stone* magazine by journalist John Colapinto (see Figure 3.1).[11] In a feature article, Colapinto reported: "For 25 years, the case of John/Joan was called a medical triumph—proof that a child's gender identity could be changed—and thousands of 'sex reassignments' were performed based on this example. But the case was a failure, the truth never reported. Now the man who grew up as a girl tells the story of his life, and a medical controversy erupts" (Colapinto 1997, 54).

In a revelation that dramatically validated essentialist claims, David Reimer told Colapinto that he was incredibly unhappy with his sex reassignment and imposed gender identity: He had never felt comfortable as a girl during his childhood and adolescence and always knew something wasn't right: "For when Joan daydreamed of an ideal future, she saw herself as a 21-year-old male with a mustache and a sports car, surrounded by admiring friends. 'He was somebody I wanted to *be*,' [Reimer] says today, reflecting on this childhood fantasy. By now Joan was ever more certain that submitting to [further vaginal surgery to complete the sex change] would lock her into a gender in which she felt increasingly trapped" (Colapinto 1997, 70). The story earned Colapinto the prestigious American Society of Magazine Editors (ASME) award. In 2000, he published a book-length account of Reimer's story, *As Nature Made Him: The Boy Who Was Raised as a Girl* (see Figure 3.2). Colapinto quotes Reimer:

> My childhood. It comes to me. I don't go and think about it. . . . Memories of how I used to look. Memories of being belittled by my classmates. Memories of just trying to survive. If I had grown up as a boy without a penis? Oh, I would still have had my problems, but they wouldn't have been compounded the way they are now. If I was raised a boy, I would have been more accepted by other people. I would have been way better off if they had just left me alone, because when I switched back over, then I had *two* problems on my hands, not just one, because of [their] trying to brainwash me into accepting myself as a girl. So you got the *psychological* thing going in your head. When I'm intimate with my wife it sometimes *haunts* me. (2000, 261–62)

Figure 3.1. *Rolling Stone* magazine, December 11, 1997.

In 2004, at the age of thirty-eight, David Reimer committed suicide.

Although David Reimer's life course seems to be strong evidence against Money's gender socialization thesis, another case suggests otherwise. As Katrina Karkazis (2008) notes in *Fixing Sex*, the John/Joan situation was one of two cases during the same time period wherein a "normal" male underwent sex reassignment because of a penis injury. However, the John/Joan case is the one that "received enormous attention" because it was the failed case: David Reimer was unhappy with his sex reassignment. In the other case, a seven-month-old baby boy was reassigned to female after his penis was destroyed in an accident (Bradley et al. 1998). At the age of twenty-six, this individual "was described as a bisexual woman with a female gender identity"—in other words, the gender socialization thesis was supported, not refuted (Karkazis 2008, 70). The latter case received little to no attention, which "points to the influence of journalism, the importance of a 'speaking' subject

Figure 3.2. The cover of John Colapinto's *New York Times* bestseller.

(Reimer), and the desire to reaffirm male-female differences as 'natural' more broadly in society" (Karkazis 2008, 70).

Medical Critiques from Intersex Activists and Feminist Scholars

As the flaws in John Money's approach to intersex medical management were being exposed professionally by Milton Diamond and other medical skeptics and publicly by journalist John Colapinto, intersex activists and feminist scholars were also critiquing intersex medicalization, specifically medically unnecessary treatments and the lies and deception that too often accompanied them. Money's gender theory legitimized the surgical interventions that ISNA founder Cheryl Chase and her supporters were organizing to end. In a 1998 *GLQ* publication, Chase theorized the movement's focus on the physical alteration of intersex bodies: "Pediatric genital surgeries literalize what might otherwise be considered a theoretical operation: the attempted production of normatively sexed bodies and gendered subjects through constitutive acts of violence. Over the last few years, however, intersex people have begun to politicize intersex identities, thus transforming intensely personal experiences of violation into collective opposition to the medical regulation of bodies that queer the foundations of heteronormative identifications and desires" (1998a, 189). Millarca, a longtime intersex activist, described the early days of intersex activism in more practical terms: "We all had a purpose, we all had a vision, and we worked hard for that vision. Our vision was to get rid of genital mutilation surgery. We wanted to get rid of genital mutilation surgery on newborn kids that were having unnecessary surgery. That was the goal of the entire intersex community and we fought hard for it."

Feminist scholars were also criticizing intersex medical care, in similar terms but within a larger context of gender structure. For example, relying on Judith Butler's (1993, [2006]1990) theoretical ideas, in 2001, sociologist Sharon Preves offered a sharp critique of the medical profession's treatment of intersex traits: "[T]he majority of these infants are medically assigned a definitive sex, undergoing surgery and hormone treatments to 'correct' their variation from the anatomies expected by the designations of female and male. The impetus to control intersexual 'deviance' stems from cultural tendencies toward gender binarism, homophobia, and fear of difference" (524). As psychologist Suzanne Kessler pointed out, "[I]n order for physicians to adopt a new stance [in approaching intersex traits], they would need to examine their own gender-related assumptions in the context of a major cultural shift in gender beliefs" (1998, 130). Preves even suggested that medical professionals have a professional responsibility to act differently: "Given the injunction 'do no harm' and the lack of data conclusively demonstrating the efficacy of more cosmetically based medical interventions, clinicians should opt for the least invasive treatment procedures and not conduct any irreversible surgical or hormonal intervention without the patient's direct consent" or, in the case of a young child, until he or she is able to contemplate all of the factors involved in the permanent decision for him- or herself (2001, 545).

In 1996, Morgan Holmes, an interdisciplinary scholar with an intersex trait, shared her personal perspective with a room full of medical professionals and gave them some advice about how to move forward in a less damaging way:

> Parents and doctors must give up ownership of the sexual future of minors. Children are no longer the property of their parents; we are not chattel. Our sexualities do not belong to the medical profession. It may be that if surgery had not happened when I was young I would have still chosen it. It is equally possible that I would have chosen to keep my big clitoris; the women I know who escaped surgery are quite grateful to have their big clits. That decision should have been mine to make. Without retaining that decision as my personal right, all other aspects of my sexual health have been severely limited. . . . The medical profession can't give back what was taken from me. But it can listen to me. I was asked to address you today from my heart, at an informal level, not primarily as an academic, but as a person who has lived through the nightmare of early childhood surgery. But I want to remind my audience that I am an academic, that I do hold a graduate degree in this area of research, and that I am a doctoral candidate specializing in this field. . . . I believe the medical profession really does want our lives to be better. Please listen to us as we tell you how to meet that goal.[12]

In voicing these critiques, intersex activists and feminist scholars (and intersex activists who are feminist scholars, like Holmes and several others) ventured a serious challenge to medical authority and jurisdiction over intersex traits.

The White Coat Gender Essentialists

John Money had been publicly and professionally exposed as a fraud by the time Dr. Peter Lee of the then-named Lawson Wilkins Pediatric Endocrine Society (LWPES) and Dr. Ieuan Hughes of the European Society of Paediatric Endocrinology (ESPE) co-organized the now infamous 2005 Chicago conference that resulted in the 2006 "Consensus Statement on Management of Intersex Disorders" (see chapter 2). It was also the heyday of intersex activism and feminist scholarship that critiqued the medical profession's treatment of intersex traits. This confluence of actions and events were challenging medical authority, apparently successfully. Dr. C. explained to me that

> anyone who ever heard the [John/Joan] story [which was often presented alongside narratives from intersex activists], physicians, especially parents, and patients [was] extremely suspicious of everything we do, and rightfully so. I mean, it was all coming out. Your integrity is the one thing that you work the longest to get and with just one slight infraction, results in total dismissal of integrity. And I'm trying to teach [my kids this] all day long, that authority is the one thing you have to guard like your jewels. . . . Now, it came under great suspicion, and I think that the only way to make it right is to make it now more clear.

The strategy for making medical authority "more clear" was to adopt *DSD*, a term that had not (yet) been politicized by activists (although Chase and her allies were secretly responsible for getting the DSD terminology on the Chicago conference agenda, those in the medical profession weren't aware of this background when they formally introduced it).

In *Deviance and Medicalization*, sociologists Peter Conrad and Joseph Schneider ([1980] 1992) introduce a five-stage model of medicalized deviance, into which, as I've argued elsewhere, the history of intersex fits nicely.[13] This model allows us to understand how "medical claims are couched in terms that attempt to conceptualize deviance as a medical problem and may be presented as a medical diagnosis or etiology and/ or treatment for the deviance and the deviant's behavior" (Conrad and Schneider [1980]1992, 266). In Stage 1, a behavior or condition is identified as deviant. In this stage, intersex was viewed as a negative deviation from the sex binary. Stage 2 is medical discovery, which for intersex involved doctors' reporting in medical journals on their discovery of intersex traits. These discoveries evolved over time, as technology and biological understanding advanced, sometimes emphasizing hormones, sometimes emphasizing chromosomes, and, at earlier stages, even focusing on social traits. In Stage 3, medical and other stakeholders make claims about the behavior or condition in question. In the case of intersex, doctors established their expertise, even during moments of medical uncertainty, by claiming that intersex was a medical problem they could fix. Meanwhile, intersex activists mobilized to change how doctors view and treat intersex. Stage 4 highlights legitimacy and securing medical turf. In the case of intersex, this stage played out as a battle over diagnoses, as doctors reasserted their authority over both the diagnosis of intersex and people with intersex traits. This leads to Stage 5, the final stage, in which the behavior or condition institutionally reaches medicalized deviance status. This is the current stage for intersex, which the widespread medical adoption of the DSD nomenclature has turned into an officially recognized "abnormality." In the rest of this chapter, I focus on this stage.

Dr. D. described the usefulness of DSD terminology, in particular the opportunity it offered to medically redefine intersex: "[*DSD*] makes clarification much easier. It removes some of the stigmatizing words. It's actually easier to teach. . . . Saying *DSD* makes it absolutely clear as a bell . . . you have an enzyme disorder in this pathway or you have a structural disorder from that pathway, this was an accidental birth defect and you were born without this part. It makes it so much easier and less threatening. . . . Physicians need taxonomic terms, and it's been tremendously helpful for medicine." Though Dr. A. was not quite as enthusiastic, he still embraced the new language: "*DSD* is a somewhat more complete and accurate term . . . *DSD* is a perfectly fine term; I now use it in my own work. I don't worry about it. Several of us wrote a paper about supporting that change [*laughs*]." Although Dr. Courtney Finlayson, a pediatric endocrinologist, was advocating on the Huffington Post blog for "acceptance and kindness" for intersex people, she still used *DSD*, knowing that her terminological choice was controversial. She explains:

> As I write this article, I worry that I will err in my wording and cause offense. In the past, terms like hermaphrodite and pseudohermaphrodite were used. These were considered pejorative, and an effort was made to adjust terminology, which led to the term DSD. Even this term, however, may brand these conditions as a disorder rather than a variation of human development. I chose to use the [abbreviation] DSD here as it is the currently accepted medical term and thus as a physician, it is the term I use. I am sensitive that even this word, chosen to refer to an embryologic process, can be upsetting.[14]

As the article was circulated on social media, I witnessed a number of intersex people criticize Dr. Finlayson on private social media sites for using *DSD*, despite the fact that she herself acknowledged the controversy over its use.

DSD nomenclature enables medical professionals to reframe intersex as a medical *disorder* that necessitates medical attention, an "embryologic process" gone wrong.[15] According to Dr. C., this change was long overdue: "First [are] the terms of the word *intersex*. Again, the word *sex* is highly emotionally charged, whether it's your gender, whether it's having sex, or anything else. And so moving away from a highly charged word like that which can mean a thousand different things to a thousand different people is what we should absolutely get away from right away . . . and I think no one would disagree . . . it's amazing it took so long to get rid of the word *intersex*."

Some doctors clearly wanted to see the shift as one that made the diagnosis less threatening, but even they used terminology like *defect*. Dr. G. explained:

> I think it's so helpful to have such a broad category. So "disorders of sex development," that's a pretty broad category, and it doesn't imply judgment, it doesn't imply that one's more severe than another, because underneath that umbrella are many, many different diagnoses. So I think parents, if they're being seen by the disorders of sex development clinic, they start appreciating that all of this has to do with how our sex develops . . . that there is genetic and hormonal factors. We often use the analogy of . . . there are many different types of heart defects. Well, our bodies are complex—how we're put together—and there are many differences in body shapes and sizes, and there are differences in how genitals have grown and developed.

Others went so far as to compare intersex traits to dangerous diseases, despite the fact that most intersex traits have minimal, if any, health risks.[16] Dr. C., for instance, viewed DSD terminology as "an analogy. It's like talking about skin cancer and brain cancer." The comparison is even clearer in a 2014 publication by David Sandberg and Tom Mazur, two well-known medical providers in the intersex community, who propose "an integrated team care model that is patient and family centered and attempts to balance the dominating perspective that focuses almost exclusively on gender-related aspects of DSD with one that *conceptualizes DSD as a congenital and chronic condition, akin to other pediatric conditions*" (93, emphasis added).

The linguistic shift from *intersex* to *DSD* served as a perfect vehicle for medical professionals to reassert their authority and maintain their exclusive jurisdiction over intersex traits. It allowed medical professionals simultaneously to move beyond the John Money debacle and to respond to intersex activism and feminist critiques that were successfully claiming intersex was not a medical problem. It no longer mattered, in short, that Money's gender theory had been exposed as neither empirically nor ethically supportable. It also didn't matter that intersex activists and feminist scholars were questioning surgical intervention. None of this mattered because medical professionals had a new group of traits to treat: disorders of sex development.

My interviews with doctors made it immediately clear that most reject Money's gender theory. In fact, seven of the ten medical experts I spoke with hold narrow essentialist beliefs about gender: They believe that gender is biologically determined, and their descriptions of it evoked Western, white, middle-class understandings of femininity and masculinity. These beliefs, in turn, had significant implications for how they justified their treatment of intersex traits. When I asked Dr. D., a well-respected endocrinologist, if gender was ever incorrectly assigned, her answer was unequivocal:

> Yes. When an individual who's been raised as a female gender assignment comes to the office having totally cut off all her hair, wearing army combat boots and fatigues . . . it sounds very stereotypical, but it really happens . . . wearing combat boots and fatigues, saying, "Oh God, I hate having periods, it doesn't make any sense for my life, I don't like this." Or they threaten to commit suicide or they're institutionalized with substance abuse, and part of what comes out of their therapy through that substance abuse is that they don't know who they are or they think they weren't assigned to the way they feel now. And those are not always permanent, by the way . . . one of my fatigues-wearing persons came in a couple weeks later, wearing a miniskirt, makeup, and having dyed her hair.

Although Dr. D. seemed to recognize that gender can be fluid, her basic position was still essentialist, founded on biology, or, as she put it, "hardwired":

> My experience with girls with [congenital adrenal hyperplasia, or CAH] suggests to me that it's pretty hardwired. A lot of the CAH girls are significant tomboys . . . hey, I was a tomboy, it's not a slap, it's a description. They're more of a risk taker, like at age 5 or 6, they'll leap off the porch because they think they can fly kind of things. They wanna take their skateboard and turn it into a hang-glider and take off from a cliff. . . . Whether it's only with DSDs, or whether it's in folks where you couldn't find a DSD with a microscope, I think some of those behaviors are absolutely hardwired.

Dr. D., like many of the medical providers I spoke with, identified masculine behaviors as risk-taking activities tied to potentially dangerous actions. This line of thinking moves beyond biological sex differences to the assumption that our gendered behaviors are biologically prescribed.

Many doctors believed that gendered behaviors resulted, at least in part, from hormonal exposure during gestation. Dr. A., an endocrinologist, said, with no hesitation: "I think there's no question, again, based upon the [CAH] experience, or the experience of kids who are exposed to androgens externally during pregnancy, that there's very good reason to believe, and there's probably experimental data about this, to suggest that androgen levels during fetal development produce male-typical behavior later on, there's just no question about that." Consistent with the binary logic that neatly correlates sex, gender, and sexuality, most doctors used these terms interchangeably when justifying their views. When I asked Dr. A. if he could clarify what he meant by "male-typical behavior," he elaborated by discussing sexuality: "Like in primates, where they've used high levels of androgens during fetal development in chromosomally female fetuses, those female monkeys are engaged in humping behaviors and things like that, which are much more typical of what male primates do as immature, and later sexually mature, individuals . . . how they engage in intercourse." Here, "humping behaviors" are associated with assumptions about males—specifically, that they are more sexually aggressive.

Medical professionals are trained to look at intersex through a biological rather than a social lens. A biological lens necessitates an essentialist view of gender to justify the medicalization of intersexuality and, more specifically, to justify the validity of the *disorder* of sex development diagnosis itself. Gender essentialism maintains that gender is—and should be—neatly correlated with a binary sex, and that the bodies of individuals who deviate from the pattern require modification for the sake of normalcy and consistency. I also believe that sexuality is a key component of gender essentialist logic, as in the field of medicine, sexuality, like gender, tends to be viewed as biologically prescribed. Most of the medical professionals I interviewed believe that intersex is typically a visible abnormality of the body that warrants and justifies treatment. When I asked Dr. D. whether a person with a DSD would be recognizable in a crowd, her response surprised me:

> Some of them, yes. Because some have some very specific phenotypic [external appearance] features . . . It's as if you're trained to look for them. [Interviewer: "And what are some of these big, obvious characteristics?"] Okay, well, Turner Syndrome. So there are phenotypic features of women with Turner Syndrome. And if you know what you're looking for . . . short stature, droopy eyes, very prominent ears, a webbed neck, there are characteristic features of the fingers, and Klinefelter's Syndrome, not always, but some of the forms of where folks are just agonadal, they have long, thin body proportions with big, long arms, and a high-pitched voice, and not much facial hair or facial musculature or shoulder musculature, you can sort of go, "That person looks like they might have a DSD."

It's not surprising, then, that several of my research participants with intersex traits expressed genuine concern that their diagnosis was visible in public settings. Yet, during my ethnographic observations at meetings of intersex social movement organizations, I found it impossible to determine who had an intersex trait and who did not (as I discuss below, a few outliers in the medical profession agree with me). As a sociologist, I view intersex through a social rather than a biological lens.

Medical professionals, with their biological lens, often rely on essentialist views about gender to justify the medical tests they use to reach definitive gender assignments—which they interchangeably call sex assignments, usage that further underscores their belief in the interchangeability of sex and gender.[17] According to Dr. Mariam Moshiri and her colleagues: "The first step in the management of DSD is sex assignment, which is based on factors such as the genotype; the presence, location, and appearance of reproductive organs; the potential for fertility; and the cultural background and beliefs of the patient's family. The primary goal of sex assignment is to achieve the greatest possible consistency between the patient's assigned sex and his or her gender identity. Once the sex is assigned, the next step in management might be surgery, hormone therapy, or no intervention at all" (2012, 1599). Involving psychiatrists only peripherally, medical professionals assign gender almost exclusively within the biomedical paradigm (a model that relies on biology to make sense of traits, without taking psychological or social factors into account).[18] As Dr. C. explained:

> We do all the biochemical information . . . we do all the morphometrics, radiologic assessment, and then we sit down—the endocrinologist, myself, sometimes a general surgeon—really, surprisingly, within our setting, very rarely, a psychiatrist. And we'll discuss primarily—in all discussions that I've had input into—who the child thinks they're going to be later. Which seems to be a fairly simple thing, which is did the child have significant testosterone exposure . . . in utero? And then once that's been established, discuss the issues such as fertility and functional success of surgery.

Framing the gender assignment recommendation within the biomedical paradigm leaves its validity unchallenged and establishes the rationalization of surgical intervention.

I found further confirmation of the medical profession's commitment to biological explanations when I asked medical professionals how they might explain gender variation. They commonly cited poor parenting, rather than question their assumptions that gender is biologically prescribed. Dr. B.'s response to a question about gender variant intersex kids was fairly typical: "I have worked with kids who've decided to kind of move forward in an opposite gender [from] the one that they were originally assigned to. But the cases that I've worked with where that happened were predominantly very poorly controlled kids with congenital adrenal hyperplasia, who were very masculinized by [that] time. I worked with one young boy, adolescent male, who was really a genetic female, had been born in another country, very poorly controlled, very masculine." Dr. B.'s response illustrates the commonly held belief among medical providers that individuals are "really" either female or male, a longstanding ideology in medicine, though it has long been subjected to critique. In his Introduction to *Herculine Barbin: Being the Recently Discovered Memoirs of a Nineteenth-Century French Hermaphrodite*, Michel Foucault (1980) problematizes binary sex and questions the reasons behind sex categorization. He argues, more specifically, that sex, understood as a biological truth, is a fabrication perpetuated by juridical and medical processes. In *Bodies in Doubt*, Elizabeth Reis (2009) offers us a specific historic account of doctors' attempting to impose their understanding of sex onto a resistant body:

> In the late nineteenth century, one thirty-four-year-old who had been living as a woman refused a recommended lengthening of her urethra and correction of what the doctors saw as a curved penis because surgery would necessitate [her] entering the hospital as male. . . . [Dr. James Little, a professor of surgery at the University of Vermont and the New York Post-Graduate Medical School] recommended the operation, most likely because he saw the patient as a man and assumed that he would want the curvature fixed. But the patient refused. . . . In part, this person's decision to forgo surgery was based on her history of living as female . . . For this particular person, considerations of practicality, convenience, and the possibility of surgical damage merged with the sheer logic, for her, of switching genders at age thirty-four. Yet, throughout his account, Dr. Little referred to the patient as male, although Little knew that she was, and planned to continue, living as female. . . . In Dr. Little's view, this patient was merely "passing" and had been doing so for years. The doctor had learned that the family suspected something was amiss years earlier. Though the child was proclaimed a girl at birth and had been reared as a girl, between the ages of twelve and fourteen, the child noticed "that he differed from other girls of his acquaintance, and calling his mother's attention to it, she consulted a physician, who, after making an examination, informed her of the nature of the deformity, and assured her that the child was a male." The parents continued to raise their child as a girl.

According to Little, they were "too ignorant to properly comprehend the difficulty . . . and in consequence made no change to his apparel." In his patronizing attitude toward the parents and patient, Little was typical of nineteenth-century doctors who saw such patients as misguided in not adhering to medical advice. The idea of mistakes would come to have profound implications. Doctors interpreted those with unusual genitals as either "really" male or female and understood patients' own reading of their external genitalia as evidence of naïve foolishness at best and willful obstinacy or perversion at worst. (75–76)

I quote this story at length because it shows multiple doctors pushing against an individual's gender preferences, and her family's support of such preferences, reflecting the practical and ideological insistence of the medical profession and an iteration of the belief that gender variation among intersex children is a result of their being "poorly controlled" by parents, a belief that gives those parents a tremendous amount of responsibility for policing their children's behaviors in gender-stereotypical ways.

Medical professionals tend to assume minimal, if any, responsibility for the consequences of their medical interventions. For example, in a 2012 long-term outcome study of six individuals born with intersex traits[19] published in *Pediatric Endocrinology Reviews*, Dr. Peter Lee and Dr. Christopher Houk conclude that outcomes "var[ied] from poor to good" (140) but shift responsibility for the less successful outcomes onto parents, although they do acknowledge other factors such as psychosocial support. In the first case they present, Lee and Houk note that "[w]hile the procedure was explained to the parents, the mother indicated later [that] she felt forced to sign the consent form because of the urgency of the situation even though she was not convinced of a female assignment" (141). The child, who was assigned female at birth and underwent "[f]eminizing surgery at 6 weeks of age" (141), decided to gender transition at the age of eleven. When he was fifteen, he "underwent a two-staged surgery to create a penis" (142). Currently, this man is living with a relative, but for some time he was involuntarily "committed into a residential psychiatric treatment facility" because of "[c]onsiderable family turmoil" (143). Despite noting that the mother "felt forced" into surgical intervention, Lee and Houk conclude that "It also seems clear that if parents of example #1 had not agreed with the reassignment to female that outcome would have been better" (149). In short, their essentialist biomedical paradigm not only allows medical professionals to justify their surgical interventions but also enables them to evade blame if those inteventions don't work out, a point I return to in chapter 5.

Troubling the Team

The 2006 "Consensus Statement on Management of Intersex Disorders" envisioned a team approach to treating individuals with intersex traits:

> Optimal care for children with DSD requires an experienced multidisciplinary team that is generally found in tertiary care centers. Ideally, the team includes pediatric subspecialists in endocrinology, surgery, and/or urology, psychology/psychiatry, gynecology, genetics, neonatology, and, if available, social work, nursing, and medical ethics. Core composition will vary according to DSD type, local resources, developmental context, and location. Ongoing communication with the family's primary care physician is essential. The team has a responsibility to educate other health care staff in the appropriate initial management of affected newborns and their families. For new patients with DSD, the team should develop a plan for clinical management with respect to diagnosis, gender assignment, and treatment options before making any recommendations. Ideally, discussions with the family are conducted by one professional with appropriate communication skills. Transitional care should be organized with the multidisciplinary team operating in an environment that includes specialists with experience in both pediatric and adult practice. Support groups can have an important role in the delivery of care to patients with DSD and their families. (Lee et al. 2006, 490)

The medical professionals I interviewed supported this recommendation. As Dr. A. put it, "I think it's pretty clear that the best care is care that includes people with expertise from endocrinology, from urology, from psychology and psychiatry,[20] from ethics, etc., etc. There are only half a dozen places in the country that have them, if that." While there are only a few established DSD teams across the United States (including Seattle, Oklahoma City, Denver, Ann Arbor, Chicago, and Cincinnati), medical providers tend to point to such centers as the gold standard of DSD care.

I'm left wondering, however, whether the multidisciplinary team becomes an uncontested space in which medical professionals can assert their expertise and enhance their authority over the intersex body, rather than disperse that authority. When medical professionals operate in a team, the assumption is that multiple opinions will be heard. Dr. C. articulated this assumption:

> We take solace in the fact that we're operating as a team; it's not generally blame, but the better way to look at that would be to say we're showing the families as clearly as possible just how much we're wrestling with the situation ourselves, and I think that's very important. . . . We generally speak among all five pediatric urologists here as a group, telling the family we've had four second opinions without even needing anybody else here. But I think we take solace in that. I think it's a very important thing for the family to see, just how much we're wrestling with the choice ourselves.

The problem with "telling the family we've had four second opinions" is that it tamps down any potential doubt parents might have about proceeding with the team's recommendations, even to the point of disempowering them from seeking their own second opinion. The existence of a team doesn't guarantee "four second opinions"—everyone on the team could easily have had the same opinion, or a vocal medical expert on the team could be dominating the decision-making process.

Furthermore, because we know that the composition of decision-making groups can affect policy outcomes,[21] including voices from outside of medicine, such as psychologists or even bioethicists or sociologists, on DSD medical teams needs to be the rule and not the exception. Above, Dr. C. explained that his DSD team makes recommendations by "sit[ting] down—the endocrinologist, myself, sometimes a general surgeon—really, surprisingly, within our setting, very rarely, a psychiatrist." A recent experimental study published in *The Journal of Sexual Medicine* showed that subjects randomly assigned to play the role of parents of an intersex child were less likely to grant consent for surgery when a psychologist presented the intersex trait in a demedicalized fashion than when an endocrinologist presented it in a medicalized fashion.[22] Imagine what the outcome might be if a sociologist or scholar of gender and sexualities studies did the presentation.

DSD teams work together to reach a diagnosis before deciding on what they believe is the best response. Dr. I. described her team's composition and process: "Number one, the child is referred to as 'baby' until we have a 'boy' or 'girl' status. . . . Once we have the [karyotype] data, we meet as a team—a pediatric urologist, a psychologist, a geneticist, an endocrinologist, and a genetic counselor, for example, that's the makeup of the team, with the possibility that we have utilized the ethics team for ethics consultation." All of the professionals at medical institutions that had teams in place described them similarly, but the fact that many teams included psychologists does not remediate their medical emphasis, for these were generally hospital-based psychologists who worked in medical environments. Consistent with gender essentialist logic, teams tend to order diagnostic tests that include assessments of hormonal levels, karyotypes, and other biological signs of sex, suggesting that gender can be uncovered biologically. According to Dr. C.: "[The team] need[s] to figure out hormonally if the child makes testosterone.[23] We need to figure out genetically what the chromosomes are and then discuss what little knowledge we have in 2010, how we think this child's going to think. Not in terms of gender preference or [whom] they're attracted to, of course, but for gender identification, who they think they are." These descriptions of how teams work can help us understand the medical profession's relatively quick implementation of DSD nomenclature. If medical professionals view gender as something that should function properly alongside sex, gender that doesn't neatly match sex is taken as a sign that an individual hasn't been correctly sexed rather than as possible proof that sex and gender are social constructions. DSD terminology aligns with this vision and gives medical professionals the authority to define and treat bodies as they see fit.

Although the medical professionals in my study were aware of the 2006 consensus statement's recommendation that surgical intervention should be avoided unless it promotes a "functional outcome rather than a strictly cosmetic appearance" (Lee et al. 2006, 491), medically unnecessary surgery has continued. Dr. A. said, "I think surgical intervention is still quite common, even in a relatively enlightened place [such as this hospital]." Although Dr. I. explained that her hospital's medical team "take[s] the decision for surgery pretty seriously," she went on to say that they still performed surgery but "probably have moved towards less surgery early on." Essentialist views of gender likely play a role in the continued prevalence of surgery, as we can see in Dr. C.'s explanation of how he presents surgery to the parents of children who are newly diagnosed with intersex traits: "I always talk about it as, nature . . . just about got it right, but just this is the last few steps or last step, and we can complete that for you, and then we take lots and lots of questions."

All of the medical professionals I interviewed stated that parents approach the intersex diagnosis with many questions, an observation supported by the parents I interviewed. The medical professionals believed that parents usually welcome their professional opinions with little resistance or hesitation. However, this is not always the case. Dr. C. recounted a recent consultation with a family that was very critical of his recommendations:

> The father said, "[Doctor], can I ask you a question?" I said, "Absolutely, this is your forum. I'm at your disposal. You're hiring me." He said, "Why should we do anything?" And I acted physically surprised, I'm sure I did. And I said, "Well, I'm concerned that if you raise this child in a male gender role without a straight penis, he's not going to see himself as most other males and he's not going to certainly be able to function as most other males." And the father said, "Well, in our family we like to celebrate our differences and not try to all be the same and feel the social pressure to do everything like everyone else does." . . . I said, "I do have to say one thing, and I think it's of key importance, that you both see a psychiatrist."

Clearly Dr. C. did not provide much space for parents to question his medical recommendations, and he was not alone. Indeed, this passage shows that, despite the 2000 and 2006 medical statements, Dr. C., like other medical professionals I interviewed, continues to be invested in surgical modifications on intersex bodies that are often cosmetic but always irreversible. As demonstrated in chapter 5, parents still tend to yield to the power and expertise of medical professionals and consent to these modifications, but later they express decisional regret.

Doctors Critiquing Doctors

While the 2000 and 2006 medical statements have not put an end to unnecessary surgical interventions on intersex bodies, they have instigated a critical discussion about the necessity of surgery. Medical professionals are not a monolithic group of individuals but hold different views about the dominant medical treatment of intersexuality. As Dr. A. explained, "Some urologists are sort of gung-ho about doing early surgical procedures, and there are others who are now very reluctant." Dr. A. went on to acknowledge that "the vast majority of surgical interventions that have been proposed or practiced in early childhood have no proven physical advantages, although there may be some arguments for the technical feasibility of doing some things earlier, rather than later, but that's a general matter." When I asked Dr. F. to share her thoughts about the discrepancy between the 2006 consensus statement and actual practices, she explained, "It's better than what existed before, but I think it didn't go as far as it could go . . . in terms of redirecting away from surgical approaches." Dr. F. also issued a plea to her professional colleagues around the country: "*Don't* do surgery . . . just leave this child as is at least until sometime later in the child's life when it's clear what this child's gender identity appears to be." In her memoir *Reclaiming My Birth Rights*, Dr. Adrienne Carmack, a urologic surgeon, had this to say about intersex medical care: "The approach by medical doctors to assign a gender, and then administer irreversible treatments to support that gender, is fundamentally flawed! No matter the original logic behind this treatment model, it is now apparent that in many cases it was a mistake. Yet, surgeries are still being done based on what is thought to be the gender a child will relate to" (2014, 67–68).

I interviewed three medical professionals (out of ten) who were critical of the binary biomedical account of sex, gender, and sexuality, but finding them required intensive purposeful recruitment strategies. Influenced by feminist scholarship, Dr. E., Dr. F., and Dr. H.[24] held very different views from those of the majority of medical experts on intersex. For example, when I asked Dr. E. if she thought gender was biologically predetermined, she grounded her response in feminist scholarship: "Well, here I'm gonna probably diverge from any biological explanations for this, 'cause I don't know if you've read any of Anne Fausto-Sterling's stuff . . . she's very convincing to me . . . gender, sexual orientation, hormones, phenotype . . . I think has to do with the way nature works, and nature loves variety . . . maybe there's some way testosterone tends to make people act more boyish. But I think it's the way we then interpret that boyishness." Dr. F. offered her own criticism of the binary logic, which didn't rely on feminist scholarship but revealed a broader social vision: "So we still have this dichotomous society that thinks in black and white, male and female, and there's nothing . . . you can't be anything but one or the other. It's some of these social constructs that seem to exist in the United States that maybe don't exist in other countries."

As noted above, the majority of the experts on intersex I interviewed believed that a person with a DSD could be visually recognized in a crowd of people, an assumption which highlights that providers are looking at intersex through a biological rather than a social lens. However, I spent hundreds of hours at intersex organizational meetings where it was impossible, with my social lens, to distinguish intersex people from their "normally bodied" parents and significant others, and the remarks of these three outliers in the medical profession support my ethnographic observations. I asked Dr. E. whether an individual with an intersex diagnosis could be easily recognized in a room full of people, and, rather than cite alleged physical markers of sex, she spoke of gender, quickly replying, "No! [*laughing*]. . . . In my experience, intersex people are so raised to conform to a gender role that they do."

Medical professionals often cited fertility as an important determinant of gender assignment. For example, when I asked Dr. F. to describe the process involved in assigning gender to individuals born with externally ambiguous genitalia, she explained:

> Basically the outward appearance. To some extent, what structures the child has internally, as well, can affect that . . . [Interviewer: "What do you mean by structures?"] Well, like a uterus for example. If a child does have part of a uterus, that can be a guide. . . . Physicians tend to go toward the female sex of rearing [female gender assignment], because that has the potential for carrying a child. This is kind of the *holy grail* of being able to bear a child and carry a pregnancy. So that does tend to drive sex rearing towards female, if there's a uterus present.

Dr. F. was the one who implored her colleagues, "*Don't* do surgery," but even she was unable to entirely escape hegemonic ideologies about gender. When I asked her whether any child's gender identity could ever be "clear," given her personal view that gender is a "social construct," she explained:

I think there's . . . yes, but I think there's plasticity in that, I think it's mal-leable. I think *we're endowed with this certain level of masculinity or femi-ninity at birth, due to whatever prenatal influences we're exposed to,* but I think there can be postnatal influences that may modify that, whether they're hormonal influences, or whether they're external, environmen-tal influences. I'm not entirely sure how environmental influences would change that, *I think it's probably more biological than sociological,* from my perspective. But I do think there are probably cases where it's malleable. And transgender individuals are kind of those examples, which are not part of this discussion. (Emphases added)

As Dr. F. illustrates, even the most progressive providers may not move entirely beyond essentialist understandings of gender.

The progressive medical professionals I interviewed acknowledged that their medical peers pressure parents to consent to medical inter-vention, particularly surgery, and to police gender. Describing one case, Dr. F. explained: "Surgeons want to do a surgery to repair the hypospa-dias and make the child look more typically male now, and the mother is concerned that if she does that, . . . the child may somehow later in life reject that, or may want to change to a female sex of rearing, and [will] have gone through all this surgery unnecessarily. So the mother is disinclined to do any surgery, and the surgeon is trying to hint her into the direction of doing surgery." Dr. E. also described parents' being pressured by medical professionals, in this case to police gender per-formance: "I think parents are really pressured . . . from doctors. Yeah, because part of the outcome was that [intersex children are] supposed to adopt that gender role. I was talking today about a [person] whose mother wouldn't let her wear tomboy clothes, wouldn't let her join the girls' softball team 'cause these were activities of men."

Dr. E. expressed real concern for intersex children. As a parent herself, she also understood the desire of parents to raise "normal" children, es-pecially because some medical professionals rely on gender conformity to confirm that the sex assignment (to which the parents consented) was made "correctly." She did her best to help parents understand that gen-dered behaviors do not necessarily correlate with sexuality: "When I talk with parents, for whom this is an issue . . . it's a somewhat easier issue now, because twenty or thirty years ago, the beliefs about what gender roles/stereotypes were, were a lot stricter than they are now; most parents now don't have trouble with their girls' or daughters' being athletes . . . but I really try to normalize that for them. There are lots of feminine, heterosexual adult women who were tomboys when they were little girls; this is not an ominous sign." Although Dr. E. framed the queer lifestyle as less desirable than the heterosexual lifestyle, she did so as a political strategy, not a value judgment. As a lesbian medical professional, she understood the heterosexism involved in the treatment of intersex children: "I think homophobia is always under this. Absolutely, in the medical community . . . and for a lot of parents there's a big anxiety . . . they don't know [whom] they're supposed to marry or have sex with. . . . That feels rough on some parts of me." She objected to homophobic views but chose to bite her tongue in order to advocate for children expressing their gender as they see fit. Because society generally does not view children as sexual beings, this approach made the most sense to her.

Although Dr. E., Dr. F., and Dr. H. are open-minded medical professionals, they are pessimistic about the possibility of change. Dr. F. believed that moving beyond binaries was "almost a dream of utopia, to think about our society even getting to that point," because "urologists would have less work [and therefore less income], so there would probably be some . . . you know. . . ." Many urologists who are experts on intersex surgically modify a child's genitalia regardless of medical necessity, if in their judgment the modification means the child will be able to fit gendered expectations more comfortably, such as the expectation that men should be able to urinate comfortably while standing. Dr. G. shared: "Some of the babies are born where the base of the penis is really where they're urinating from. If the baby is going to be raised as a girl, that's an okay place. But if they're gonna be raised as a boy, then it may be that they're really needing to create the urethra tube and have the urine come out of the penis's tip. [But] some of the surgeries that are done on older children, you really get terrible outcomes." Dr. F. was critical of such surgical intervention based solely on social norms: "Why do [men] have to be able to urinate standing up? What's wrong with sitting down? Women sit down to urinate, so why can't a boy sit down to urinate? There's nothing physically wrong with sitting down to urinate."

Professionals like Dr. E., Dr. F., and Dr. H. are rare in the medical world of intersexuality. They disagree with the norms held by their peers because they have been influenced by feminist scholarship and understand the social construction of gender. However, although almost all of the providers I interviewed had been exposed to feminist scholarship, most had not been influenced by it. In fact, one medical professional went so far as to joke that "feminist scholarship is great for a rage-filled feminist agenda . . . in medical practice, not so much."

Conclusion

The medical profession is a powerful institution in the gender structure, which makes the medical management of intersex traits a unique access point for the study of diagnostic processes, specifically the process of naming. In 2006, *DSD* was formally introduced by the medical profession. By 2010, it had come to replace intersex language in virtually all corners of medicine.[25] Today, in the United States, medical experts on intersex traits rarely use intersex language.[26] Why did they so quickly abandon intersex language in favor of DSD terminology? We might assume that the 2006 consensus statement had something to do with it, because it both introduced the new nomenclature and was produced by an influential consortium of experts on intersex from around the world. Yet even as the statement's recommendations for nomenclature and medical management teams have been implemented successfully, other recommendations, such as avoiding medically unnecessary surgical interventions, have been far less successful, making the statement itself an inadequate explanation.

In this chapter, I have argued that when *DSD* was introduced, medical authority and jurisdiction over intersex traits were in jeopardy, as a consequence of both the John Money debacle and criticism from intersex activists and feminist scholars. Intersex, in short, was being framed as a social rather than a medical problem, which in turn was fracturing the institutional level of gender structure as reinforced by the medical profession. DSD terminology offered a unique opportunity for medical professionals, now working in supposedly well-intentioned medical teams, to reassert their authority, reclaim their jurisdiction over intersexuality, and reaffirm the gender structure.

DSD nomenclature reifies the essentialist understandings of sex, gender, and sexuality held, as I've shown here, by many medical experts on intersex traits. It does so by constructing sex as a binary phenomenon explained by science, which in turn increases the credibility of the medical profession. At a crucial moment of potential transformation, the *DSD* framework enabled medical experts to bring intersexuality back into the domain of science and place it neatly on medical turf, where surgery, among other treatments, could be justified against its activist and feminist critics. Although some providers today seem to be more careful about genital surgeries, at least in writing, acknowledging, for example, that "one should [proceed] cautiously before considering irreversible surgery," there is still "no agreement regarding the criteria for early or late genital surgery, particularly vaginoplasty," suggesting that surgery itself continues to hold sway (Lee and Houk 2013, 4–6). Medical professionals, particularly surgeons, appear to have reclaimed their authority over the intersex body and, in the process, their capacity to perpetuate ideologies which maintain that sex, gender, and sexuality are essentialist, binary, and correlated characteristics. However, a handful of the medical professionals I interviewed were critical of these narrow ideas about sex, gender, and sexuality held by their peers, which is potentially promising for the medical management of intersex traits and, more generally, the instability of the institutional level of gender structure. Surely, the medical management of intersexuality would be very different today if challenges to essentialist ideas about sex, gender, and sexuality were the norm rather than the exception.

One of the other "successes" of the 2006 "Consensus Statement on Management of Intersex Disorders" has been the implementation of multidisciplinary medical teams, which are surfacing at medical centers across the country. While this model can disperse medical decision-making power, it also can reassert the authority of medical professionals over intersex traits by vesting it in a group, not simply an individual. For medical management teams to realize their potential for dispersing decision-making power, powerful medical professionals need to be willing to give up some of their authority. If we genuinely want to improve the lives of individuals who are affected by intersexuality, we need not only to ensure that all voices on the team are taken seriously and given equal weight but also to reexamine the makeup of the teams themselves. Although DSD teams currently operate with expertise from across medicine—for example, surgery, endocrinology, and urology—

they are less likely to include psychiatrists and social workers, who may be excluded because they are more likely than surgeons to reject essentialist beliefs about sex, gender, and sexuality, let alone gender scholars from sociology, anthropology, and/or bioethics or individuals with intersex traits and their parents. Research shows that medical providers, patient advocates, and parents hold significantly different perceptions of intersex;[27] ideally DSD teams would represent all of these stakeholders.

The medical management of intersex traits clearly demonstrates that diagnoses are defined through other social constructions. In the case of intersex, the dominant construction has been the biomedical paradigm with its essentialist—and deeply problematic—understanding of sex, gender, and sexuality as binary and correlated characteristics. Expanding DSD teams to include, at the very least, a gender expert taken seriously by other team members would likely lead to different recommendations to parents of intersex children, especially pertaining to surgical intervention. Changing medical policy and practice will be a long-term endeavor, as the progressive medical professionals suggested, but this is one concrete step in that direction. The medical profession may perceive it as yet another challenge to their authority and, ultimately, their jurisdiction over intersexuality. Yet if these teams continue to ignore the gender structure and its pernicious consequences, the medical management of intersexuality will remain exclusively on medical turf, and intersex people will continue to face surgical intervention, gender policing, and unnecessarily stigmatized lives.

4

The Power in a Name

It was July 2010, and I was staying at a hotel not far from Nashville's Music Row. Although I wouldn't publicly identify myself as a country music fan, I was secretly excited that Nashville had been chosen for the 2010 AISSG-USA annual meeting. Although I had never been there, I'd always wanted to visit Music City after seeing it featured on a Travel Channel program. But I never got to see much of Nashville. In fact, I rarely left the hotel during the four days I was there. Each day was filled with back-to-back sessions, ranging from presentations of treatment options for osteoporosis[1] to intimate discussions of personal experiences with intersexuality.

I still recall one session about disclosure. Given that most intersex people keep their diagnoses to themselves or share them with only a selective few, I figured it would be a well-attended session, and it was. The room was so packed that several participants found themselves sitting on the floor, like elementary school children. Of the approximately forty people I counted in the room, most were women with intersex traits, although there were a few researchers, including two well-known sociocultural scholars. Two outspoken activists who had lots of experience publicly sharing their personal histories with intersex facilitated the session. One had recently appeared on NPR, and the other had been quoted in numerous newspapers and appeared in several new documentaries. They began the session by telling their stories of going public with their intersex diagnoses. However, a debate about DSD nomenclature quickly surfaced.

It had been only a few years since the medical profession had advocated for a shift to DSD terminology in their 2006 "Consensus Statement on Management of Intersex Disorders" (Houk et al. 2006; Hughes et al. 2006; Lee et al. 2006). The consensus statement helped medical professionals reclaim jurisdiction over intersexuality and the intersex body (see chapter 3), but it also created a tremendous amount of new tension in the intersex community. A few minutes after general discussion began, one of the session attendees directly asked the intersex people in the room why they preferred DSD language to the terms *intersex*, *intersexuality*, and *intersexual*. Supporters of DSD language explained that DSD nomenclature was useful medical terminology based on how bodies develop during gestation, whereas intersex was a politicized identity. Perhaps not surprisingly, the woman who asked the question was an intersex activist who had been involved with ISNA from early on and had appeared in one of its first documentaries. Yet, now that she wanted to retain intersex terminology, she had become an outlier among this group of her peers.

The DSD nomenclature debate extends far beyond this particular conference session. Since the 2006 consensus statement, it seems people in the community have had to choose whether to use intersex language or DSD terminology. The premise of DSD terminology is that framing intersex bodies as having a disorder of sex development defines them as nothing more or less than a biological phenomenon, although the word *disorder* points to that phenomenon as abnormal. Those who favor the new diagnostic terminology claim that DSD language makes it easier for the public to understand intersex, with the result of fewer stigmatizing outcomes. They also say that DSD language helps parents understand their children's intersex traits. Still, although only a few individuals at the 2010 AISSG-USA conference rejected DSD terminology, I subsequently found among other groups in the intersex community many more who refused to embrace it.

In chapters 2 and 3, I examined how intersex operates at the institutional levels of gender structure. In this chapter, I provide a systematic analysis at the individual level of gender structure in order to make sense of the diverse reactions to DSD nomenclature I have observed in the intersex community,[2] as well as to explore the connections between terminology preferences and access to biological citizenship. Influenced by Michel Foucault's (1978) discussion of biopower (power over bodies), Nikolas Rose and Carlos Novas (2005; see also Rose 2007, 2001) developed their concept of biological citizenship, which, they contend, is both collectivizing (e.g., rights, digital, and informational biocitizenship) and

individualizing.³ Placing gender structure theory in dialogue with biological citizenship allows us to see that access to biological citizenship
is related to ideas about gender.⁴ Specifically, this chapter asks: What
emotional and physical struggles do those with intersex traits face in
contemporary U.S. society? How, if at all, do intersex people respond to
their struggles? How do specific individuals and the intersex community
as a whole feel about DSD nomenclature? How, if at all, do these terminology preferences map onto access to biological citizenship? More
specifically, are all intersex people allowed similar access to biological
citizenship? Last, what does biological citizenship offer intersex people,
and what does it seem to limit?

Given that naming is such a politicized and controversial topic in the
intersex community, I feel it is important to disclose my own preference.
My research has led me to see utility in both intersex language and DSD
terminology. As I traveled around the United States, meeting and speaking with intersex people, I found that each term comes with advantages
and disadvantages. For example, *disorder of sex development* can be
pathologizing in emotionally harmful ways, but it also affords benefits
through the biological citizenship it allows. Many of the people I interviewed who embraced intersex language have positive self-identities but
also have troubled relationships with medical professionals and parents.

At the same time, those who embraced DSD nomenclature tended
to report positive relationships with medical professionals and family
members. This is undeniably a positive pattern, but we need to ask at
what cost this benefit surfaces. *Disorder of sex development* implies that
the sex binary system is natural, and the context in which it is deployed
suggests that sex is correlated with gender and sexuality, which might
explain why those who prefer DSD terminology tended to question
their gender authenticity. I embrace the notion that gender is a socially
constructed phenomenon—that what we deem masculine or feminine
depends on the context, time, and space of the individual deeming it
so—which allows me to accept my intersex trait as a normal sex variation that is not tied to my gender identity.

These observations frame my contention that, as a community, we
ought to recognize the utility of both intersex language and DSD terminology. To do so, however, we must reclaim the power embedded
within nomenclature, which, in turn, we can begin to accomplish only if

we acknowledge that, like sex, gender, and sexuality, medical diagnoses are only as real as their definitions. Underlying this chapter, then, is the crucial question: If we share the goal of reducing the shame, stigma, and secrecy associated with our bodies, why are we giving terminology the power to create further rifts and divides in *our* community?

Intersex in a Binary World

A child born with an intersex trait encounters a world that narrowly and inaccurately assumes there are only two sexes, male and female, and considers any deviation from this two-sex model to be abnormal.[5] In other words, biological sex is commonly thought of as a binary phenomenon, in which bodies are either male or female, with distinct differences between the two. However, this line of thinking is flawed, for it assumes both that sex can be captured and that we have reliable markers of sex, when we simply do not. Throughout history, medical scientists have attempted to define sex using external and internal genitalia, sex hormones, chromosomes, and even the brain.[6] Societal institutions reinforce this instinct to define. For example, birth certificates almost always require that babies be labeled male or female.[7] Of course, the sex binary has been challenged by a range of voices, from queer activists to feminist scholars and even a few progressive medical professionals.[8] However, despite such challenges, the ideologies that underline the sex binary remain influential, not only in the medical approach to intersex traits (especially surgical interventions aimed at "normalizing" the body) but also with regard to how people—especially intersex people—understand their bodies, gender, and sexualities in ways that can be emotionally harmful.

As described in earlier chapters, medical professionals have long modified intersex bodies surgically in an attempt to help intersex people fit more comfortably into the sex binary. Consistent with previous research (e.g., Karkazis 2008; Preves 2003), thirty-three of the thirty-seven intersex people I interviewed underwent surgical modification (89 percent), leaving them emotionally and physically scarred.[9] The emotional consequences of surgery mostly resulted from parental lies and doctors who attempted to keep diagnoses a secret. Ana, a thirty-seven-year-old woman, described her experience with clitoral reduction surgery:

When I was 12 . . . I was told [by my parents and doctors] that my ovaries had not formed correctly and that there was a risk of cancer and that they needed to be removed. And I had lots of examinations including of my genitals, but I was never made aware [prior to surgery] that anything was going to be taken away [from my external genitalia]. So it was a big shock to me [when I woke up after surgery]. And I really had some work to do when I was eventually ready to do it . . . from the trauma that I had from waking up from my surgery to realizing that what was between my legs was gone.

Eve, a twenty-two-year-old woman, told me that her intersex diagnosis was kept from her until she was in middle school. When her mother finally told her, Eve felt upset that this information had been withheld from her for so long. She also wondered what the diagnosis meant in terms of her identity. She explained:

I was really confused and I was angry. And I was really sad and shocked at finding out this thing about my body that I didn't know. I remember feeling very confused just about who I was after that point. . . . [It] would prove to be something that I struggled with for a long time, just my identity overall. Not just about AIS [androgen insensitivity syndrome], but like who am I because everything I thought I was . . . just was gone, I felt. It just had all crumbled. I feel like a lot of my adolescence and my teenage years were just kind of working up to that point, like "Who am I?"

When Eve was younger, she underwent the surgical removal of her testes (sometimes called a gonadectomy), which she wishes had not happened:

So [my parents] talked to doctors and I guess the only really pressing thing was the doctor said like "Okay, well when she's still really young, you have to get the gonadectomy because there's risk of cancer." And now, we know that it's not as much of a risk. . . . I have a huge scar [on my abdomen] and [my clothes] always fall down to the scar and . . . I don't know . . . obviously I had no control over it, but I regret that I had to have that surgery when I was really young.

Prior to the 2000 "Evaluation of the Newborn with Developmental Anomalies of the External Genitalia" statement (Committee 2000), which advised against such deception, intersex people, like Ana and Eve, were often lied to about their diagnoses because doctors were concerned that knowing they had an intersex trait might disrupt children's gender identity formation. While this deception is far less common today, which may reduce the emotional consequences of surgery, intersex genital surgeries continue.[10]

Even if the emotional effects are mitigated, surgical interventions aimed at helping intersex people can result in negative physical consequences. Twenty-three-year-old Pidgeon[11] said, "My clitoris is gone . . . My vagina looks really fucked up . . . There's some scar tissue there and . . . penetration hurts." Only four of the thirty-seven intersex people I interviewed were not surgically modified (11 percent). Given the physical consequences of surgery, it is no surprise that only one of them was interested in such treatment. Although Pidgeon (who prefers the gender pronoun *they*) indicated that they would ultimately support any adult with an intersex trait who chose to have genital surgery, they would still passionately advise against it: "[N]ever let them touch you in terms of surgery. That's number one. If they ask about surgery or ask your opinion, don't do that. Don't do surgery, no matter what they say. . . . You'll love your body somehow, someway, and you don't need surgery to love your body and love yourself. . . . If you fuck with your body, you can never change that. But if you don't fuck with your body, you can change your acceptance of your body." As Anne Tamar-Mattis (2006), a prominent advocate for the intersex community, puts it, "One thing that is clear about genital-normalizing surgery is that it does not consistently accomplish its apparent goals; in fact, it sometimes causes the problems it purports to solve" (72). More recently, Milton Diamond, a well-known critic of the John Money model of treating intersex (see chapter 3), and his co-author called for "a moratorium on early surgical intervention" because "the best ethical and scientific considerations require that gender surgery should be delayed until the child can consent" (2014, 2).

Given the consequences of these irreversible surgical interventions, it behooves us to ask why they continue.[12] Despite minimal evidence, many medical experts on intersex traits still claim that surgery minimizes cancer risks.[13] A more plausible explanation has to do with the binary sex model that many medical professionals—like much of society—believe should neatly map onto gender and sexual binaries.

Fifty-three-year-old Donna was told by her doctor, "You can go to college . . . have sex with any boy you want to." As she described it, "He pushed me to be feminine, he pushed me to be heterosexual, he pushed me to give in to boys." Pidgeon had a similar experience as a preteen in the 1990s, when a doctor asked, "Wouldn't you like to have normal sex with your husband when you're older? We can just fix [your vagina], we can make it a little bit bigger for you. It's just a little snip incision, and [your vagina] will just be a little bit bigger. And then you can feel like normal women and have normal sex." This line of thinking rests on the dangerous assumption that people should want to fit into the sex binary at any cost, including medically unnecessary surgical intervention. It also assumes that heterosexual partnering is the only route to "normalcy," whereas intersex people, like the rest of society, have diverse sexual identities. Out of the thirty-seven intersex people I interviewed, 32 percent identified as "straight" or "heterosexual." However, almost as many identified as lesbian, gay, or homosexual (30 percent). The rest identified themselves as bisexual (11 percent), queer (11 percent), asexual (8 percent), or reported that their sexuality was, in more or fewer words, "complicated" (8 percent).

Many intersex people, including those who underwent surgical interventions, did have a tremendous amount of anxiety about their "abnormalities," which made it difficult, if not impossible, for them to form sexually intimate relationships. But, as I first argued in 2014,[14] surgery is not the answer to overcoming those struggles.[15] Aimee, a thirty-year-old woman who had surgery, told me that the very thought of a romantic relationship resulted in a "crippling effect of fear." Mariela, a twenty-nine-year-old woman who also had surgery, commented, "I'm still really self-conscious about my body . . . and I'm worried about falling in love and when to disclose. . . . What if . . . he decides he doesn't want to be with me anymore?" Skywalker, a thirty-seven-year-old woman with a surgical history, had similar concerns, although they receded after she found a partner. She explained, "I'm enough of a woman that he doesn't care and that's enough for us." Stevie, a forty-four-year-old woman who had surgery as a young child, explained how the silencing of her experience affected her ability to relate to others: "So not only was I wounded physically through surgery . . . which I still am dealing with and may surgically revise at some point. . . . I was wounded by the mantling of my very existence being something that should not be discussed . . . the whole notion of connecting with other people especially in intimate relationships." If the goal of surgery is to facilitate a sense of normalcy, in large part for the benefit of intimate relationships, it clearly did not work for these people.

If surgery does not overcome the shame, stigma, and anxiety experienced by so many in the community, it is also not their exclusive cause. Many people shared the struggles described above, regardless of their surgical history. Kimberly, a thirty-eight-year-old woman who did not have surgery, said, "In relationships I have had, it's always really bothered me that I feel I have to disclose that I'm intersex before I get physically intimate with anybody because it's not like I could fake being normal. . . . I've always resented that." Caitlin, twenty-six, another woman who did not have surgery, echoed this sentiment: "Being in sexual relationships with people, that was really hard. I didn't have a positive relationship with my sexuality when I wasn't being honest with the people I was sleeping with."

People respond in different ways to these feelings of abnormality. At the institutional level, some turn to activism, critiquing medical treatment and raising public awareness about intersex traits. At the individual level, responses vary from avoiding sexual intimacy to seeking out normalization through heterosexual encounters. Some people rejected the medicalization of their bodies altogether, claiming intersex as an identity. Although there was no need to confine themselves to a particular strategy, most chose one and did not shift from it.

The first and perhaps most disheartening response involved avoiding sexual intimacy. Emily, a fifty-year-old woman, explained, "I don't do intimacy very well. . . . Somewhere around thirty [years old] or so I decided to screw it. I might just not even bother anymore. . . . I'll just be single. . . . That's worked out better." Marilyn, another fifty-year-old woman, said, "I still haven't had sex. . . . I haven't had a date in the last twenty years. Not a single date." Aimee also avoided intimacy, an aversion that even her therapist could not help her overcome. She reported that her therapist would say, "I don't really know how to help you with this, other than to tell you: you need to just get out there and get some experience and get over it." Avoiding sexual intimacy seemed a reliable response to the struggles associated with intersex because it enabled people to avoid their fears of abnormality. However, it was also limited, for it did not confront the struggles associated with intersex traits but rather left them intact.

The second way people responded to these struggles was to focus on heterosexual encounters that would validate their assigned sex (on the assumption that sex correlates neatly and biologically with gender and sexuality). Many medical professionals encouraged this response by urging people to pursue heterosexual relationships. Although fifty-two-year-old Ann didn't respond to her struggles in this manner, she recalled feeling that her lesbian sexuality would be a problem for her endocrinologist. She explained, "I remember him asking me if I was . . . after the surgeries were done . . . if I was dating boys. . . . [I]n my mind . . . the right thing would be to say 'Yes, I am.' . . . I remember thinking that I should just tell him that I [was] even though I was not."

For medical providers, heterosexuality has always been an important factor in determining the success of intersex treatment (see, for example, Reis 2009), and it apparently remained so for many of the people I interviewed. Tara, a twenty-three-year-old woman, said, "I slept with a decent amount of guys that I . . . don't think I should have, but I think it was the whole fact that I wanted to feel like a . . . a woman." Leigh, a twenty-four-year-old woman, had a similar experience: "When I was a young adult . . . like sixteen through . . . twenty, I went through this period where I was trying to prove to myself that I was feminine and I just engaged in some risky sexual behaviors [with men]. I think that really interfered with my life, and left some lasting marks." Jenna, a thirty-one-year-old woman, also relied on heterosexual encounters to normalize her body and feel appropriately gendered:

> I still have issues with the fact that I have AIS [androgen insensitivity syndrome]. . . . It's not a debilitating sort of thing . . . but . . . I think about it frequently. . . . I don't feel like I'm less of a woman or anything . . . but one thing I have noticed about me, like sexually, is that . . . I have something to prove. . . . I'm like . . . I'm a woman damn it! . . . And I'm going to take care of business and . . . you're gonna be like: . . . "That's the best I ever had!" . . . It's like on some sort of subconscious level . . . I want to prove that I'm a woman and I can take care of this man's needs.

When I asked Jenna to elaborate, she provided an example:

> Let's say that your orgasm is a 100 on a scale of 0 to 100 . . . for me having my partner reach climax which obviously with a dude it's ridiculously easy . . . but having my partner climax is 95 out of 100. . . . It doesn't make me . . . but the satisfaction I get from that . . . is almost as much as me orgasm-ing . . . 'cause I'm like: FUCK YEAH! I DID THAT!!! THIS XY!! BA BAM! I'm not joking, that's how I am! . . . I'm like THAT'S WHAT I'M TALKING ABOUT!

Tara, Leigh, and Jenna all sought out heterosexual activity to normalize their intersex trait and feel, as Leigh put it, "feminine."

The third response to the struggles associated with intersex was to reject medicalization by embracing intersex as a component of one's identity. Caitlin described what the term *intersex* meant to her: "I feel very emotionally connected to that word because it really did change my life for the better. So kind of moving away from that word does definitely bring up some emotional response of like no! [*Intersex*] really empowered me." Those who employed this response overcame their struggles without relying on relationships with others, something they wished everyone with intersex traits could experience. Consequently, they were often critical of those who looked to relationships to feel normal. Millarca, a forty-six-year-old woman, explained: "These girls are in relationships because they're trying to be normal. They don't want to be different, but they are different and they can't accept that. We're different. You're different, and I'm different. That's where the turmoil lies, in trying to be something you're not. If you can accept who you are, like I have, like other people have, what other people say don't mean shit. You're not trying to switch into some other box where you know damn well you can't fit into." For Millarca and others who chose this path, the first steps in overcoming the struggles associated with intersex were to embrace the intersex trait and reject the search for normalcy.

Although Millarca was critical of those "trying to be normal," she was also sympathetic to their situations and hoped they would eventually come around. For many, accepting their intersex trait was a process that started with information. Irene, a seventy-two-year-old woman, said, "I grew up being heterosexual, and . . . I think I evolved as I found out my condition. I felt more in touch with both sides of how I feel and so I feel I'm somewhere in between." Leigh, who earlier in her life had responded to her feelings of abnormality by seeking out heterosexual encounters, decided in her twenties to embrace her intersex trait and reject the two-sex model. Although she remained interested in men, by the time I interviewed her she was identifying as queer, and the struggles she'd experienced as a teenager had significantly subsided. Similarly, as a teenager Ana adopted the first response, avoiding intimacy, by burying "any hint of sexuality at all." Now, as a partnered woman in her late thirties, she embraces her sexuality, which she understands as more fluid. She revealed, "I probably could have gone either way. But I probably decided at one point that women are for me and I'll put my energy there and I won't think about it anymore." Kimberly had a similar experience: "Growing up, and dealing with being intersexed and ashamed and the secrecy, I was very asexual. I simply didn't allow myself to have an orientation. I didn't allow myself to be attracted to anybody. Now I can be, and I'm really enjoying it. I think [my female partner's right], I never wore dresses before because I never felt I could pull them off. Now I feel like I can. I've got enough ego to really enjoy the body [*laughing*]."

Rejecting the idea that intersex was an abnormality resulted in a strong sense of liberation that was often enforced by supportive friendship networks. Leigh said, "My friends have been incredibly supportive and really love me for this. One of my friends actually made me a t-shirt for my birthday one year that said, 'She's bringing intersexy back.'" Rejecting medicalization allowed Kimberly and Leigh to move from a place of shame about being differently bodied to acceptance and uplifting humor. These were among the most positive outcomes in my study.

DSD Divides

The renaming of *intersex* as *DSD*, described in chapters 2 and 3, received mixed support from people in the community, with most being against it. For example, Hida Viloria, chairperson of Organisation Intersex International (OII) and director of OII-USA, wrote in an op-ed: "I respect everyone's right to identify as whatever they want to, but personally, saying that intersex is what I am feels much better than saying it's something I have—like a disease. I can see why describing us that way works for medical practitioners invested in providing us with medically unnecessary treatments, but I don't think it works well as a means to being treated as equal human beings who don't need to be 'fixed.'"[16] Tony Briffa, a leader in Australian intersex activism, shared Hida's views: "I am personally against the use of 'DSD.' That very term turns intersex variations into diseases requiring medical intervention, and [its] being a 'disorder' inherently puts the medical profession in the leading position as experts over intersex people."[17]

What my interviews with intersex people revealed is that terminological preferences at the time[18] of the interview tended to be related to whether one understood gender as a socially constructed phenomenon or as an essentialist characteristic of the body. The majority of intersex people I spoke with believed that DSD terminology was pathologizing and spoke negatively about it (62 percent). These people also tended to see gender as a socially constructed phenomenon. As Kimberly explained, "I love the term *intersex*. For me, it really truly describes me. I am somewhere in between. I believe there's a continuum; it's not a dichotomy." Kimberly also rejected the dichotomous view of gender and understood that gender was not a neatly prescribed biological phenomenon: "I really do believe that most of the differences between men and women are not nature, as much as most people seem to think so. Not that I'm going to go [for] the 100 percent 'nurture' kind of thing, but I'd probably go [for a] 60 [or] 65 kind of split. I do think there's a lot of a difference, but, again, I do think that as a society we really enforce [those differences]."

Pidgeon had an enthusiastic preference for "*hermaphrodite* or *intersex* [terminology] . . . I feel like the language shift to *DSD* makes no sense to me . . . I don't feel it was necessary." Indeed, Pidgeon saw gender as something to have fun with: "Play with your gender if you want . . . [*laughing*] . . . you can do whatever you want! It's like, don't just stick with what you've been assigned, check out all avenues of sexuality and gender and have fun with it. See it as a positive thing, because it is a positive thing. Being a hermaphrodite is so cool! And get a hermaphrodite tattoo if you want! [*laughing*]."[19] David, a sixty-two-year-old intersex man who said he didn't "give a flying fuck" about DSD terminology and instead preferred intersex language, had a similarly playful understanding of gender: "When I was in my twenties, I played around a lot with gender fucking and just loved it." When I asked how he defined gender fucking, he replied, "Well, just fucking with your gender. I used to like to do drag and wear makeup." Stevie, the forty-four-year-old woman we heard from earlier, also believed that gender was a performance that could be altered, for instance, by using lipstick if she so desired: "Ultimately when we look in the mirror . . . and we're like either shocked by, oh my God, I need some lipstick . . . or oh my God, I want to toughen up . . . look more macho or butch . . . we basically are responding to the inner conversation in our mind's eye of what we want to see . . . how we want to appear . . . how we want to be perceived."

Like Pidgeon, Stevie was critical of DSD terminology: "For me, as an intersex person, I think that is a strong identity to relate to. I'll always identify as intersex. I will *never* refer to myself as a person with DSD." Jeanne, a forty-nine-year-old woman with an intersex trait, explained that "*disorders of sex development* is such a mouthful . . . and it is kind of a cold word . . . [but] *intersex* . . . I identify with it." She also held a socially constructed understanding of gender. When I asked her whether she felt more masculine or feminine, she said, "Probably more, maybe just a little more, feminine than masculine . . . maybe in the way I choose to present myself . . . by wearing makeup or dresses, you know, the conservative way to cross your legs." Millarca expressed similar discontent with the DSD language, emphatically stating that "*DSD* is not . . . is not something a lot of people want to identify with . . . nobody wants to be a disorder . . . who wants to be a fucking disorder? . . . I don't." She too understood gender as something that one does: "I know femmes that wear makeup and that pluck their eyebrows and pencil them back in, which I don't understand. They walk different and they dress more feminine."

When I directly asked Rebecca, a thirty-five-year-old intersex woman critical of DSD terminology, how she believed boys and men differ from girls and women, she immediately replied, "There is not a high degree of difference to be honest . . . outside of things like strength and height . . . the overlap between the two gender roles is so huge to almost make the differences between the two averages statistically insignificant. . . . In terms of behaviors . . . really, honestly . . . no." Similarly, Donna, who favored *intersex* over *DSD* and was introduced earlier, said, "Look . . . neither sex nor gender [is a] binary system . . . neither [is] black or white. . . . nature is not perfect . . . gender is socially constructed . . . what clothes you put on, how you cut your hair, how you present, how you choose to identify is gender."

In most cases, people developed these socially constructed views about gender through reading gender scholarship, in classes or on their own. Eve explained: "I've been thinking about like 'hmmm . . . do I feel feminine?' Well yeah, I do. Especially after reading Judith Butler, it's like I'm so confused about the difference between sex and gender! [*laughing*]. She confuses me so much, but I know that I feel comfortable in my feminine body, probably because I was raised a girl by my parents." In *Gender Trouble*, Judith Butler ([2006]1990) explains that "gender is an identity tenuously constituted in time, instituted in an exterior space through a *stylized repetition of acts* . . . . [B]odily gestures, movements, and styles of various kinds constitute the illusion of an abiding gendered self" (191). Chris, a fifty-seven-year-old man with an intersex trait, was similarly influenced by Butler: "I've learned a lot from [Judith Butler] . . . especially her theories surrounding [the idea that] gender is performative . . . that affected me a lot. But I find her harder to read. . . . [Doctors] are not assigning sex on birth certificates; they are giving gender identities. That's why Butler and I both agree that gender is performative; it's based on performative discourse. [Whom] you can marry is based on performative discourse, which all derives from this original discourse over your body." Aimee had a similar experience after reading work by Anne Fausto-Sterling, who has written extensively on intersexuality with the goal of showing how the narrow understanding of sex as a binary and natural phenomenon is a consequence of the scientific politicization of bodies. Aimee explained:

I didn't know the difference between sex and gender until I was a senior in college . . . and honestly, for me, that was really freeing for me . . . because honestly I felt wrong saying that I felt like . . . I felt like a liar . . . saying that I was a woman . . . and then I learned that sex and gender are two different things. . . . And even if I do have XY sex chromosomes I identify as a woman . . . that was really powerful for me to learn that distinction. [Interviewer: "Do you remember what you've read?"] Yeah, we read a piece by Anne Fausto-Sterling . . .

But not everyone's path to understanding gender as a social construction was through feminist scholarship. Caitlin, an intersex woman introduced earlier who was critical of disorder of sex language, embraced a socially constructed view of gender after working as "a stripper for a few years." Stripping, she said "was a performative femininity, and that kind of started my femme identity in a lot of ways. Like it was a conscious, performed femininity so it started to be kind of fun to play with and it started to kind of seep out into the way I presented myself." Regardless of their path to conceptualizing gender as a socially constructed phenomenon, those who were critical of DSD language tended to see gender as a performance.

Describing gender as a socially constructed phenomenon or performance did not necessarily mean debunking traditional displays of femininity or masculinity. During my two-part interview with Chris, he was "doing gender"[20] in a stereotypically masculine fashion, sporting a full salt-and-pepper beard and wearing masculine clothing. Jeanne, Donna, and others were also "doing gender"[21] but in a stereotypically feminine manner. The key factor that distinguished the many intersex people I encountered during my data collection who saw gender as a social construction but performed gender in normative ways was that they recognized that they had control over their gender performances. That is, they rejected the idea of gender as an essentialist characteristic they could not control.

In contrast to those who embraced gender as a performance, 32 percent of the intersex people I spoke with expressed an anxiety that they were not authentically gendered.[22] In many cases, these feelings of abnormality were connected to the belief that everyone has a true sex, male *or* female, that is expressed through gender. People who felt this way tended to welcome DSD nomenclature, like Tara, who stated, "*Hermaphroditism* and all [related terms like *male pseudohermaphrodite*] . . . I am not a fan of, obviously." She went on: "After I found out that I technically am a genetic male . . . when I wear a baseball hat or something I kinda look in the mirror and I'm like, do I look like a dude? . . . Some women obviously look like women." Tara's understanding of gender is grounded in gender essentialism—that is, the assumption that people are born into a correlated sex and gender.

Marilyn, a fifty-year-old woman with an intersex trait, noted that intersex terminology "bothered me a little bit because it was just a little bit too political." She also struggled with whether she was authentically gendered: "When I was growing up, I was having a hard time feeling very feminine because I wasn't developing . . . I didn't feel like a complete woman." She went on to explain that her feelings of not being "a complete woman" had diminished as she got older; but she concluded, "I still don't feel like a complete woman." Karen, a fifty-two-year-old woman with an intersex trait, also touched on the gender politics when I asked her about her terminological preference. At the time of the interview, she favored DSD nomenclature and considered intersex terminology to be "bad because it describes a possible third sex or worse . . . a limbo state between them, and I don't think humans are in limbo."

When I asked Vanessa, a forty-three-year-old woman with an intersex trait, which terminology she preferred,[23] she said, "*Intersex* rubs me the wrong way . . . I'm comfortable . . . with *disorder of sex development*. It's the development of your sex in utero . . . I think it explains . . . something that happened versus something that you chose." She also articulated a more essentialist understanding of gender: "When you think about kids, boys can be a little more physical, a little bit more aggressive, a little bit more active . . . even though girls can be [physical and aggressive]." Upon further probing, Vanessa indicated that she believed these differences occur naturally.

Although Hannah, a forty-two-year-old woman with an intersex trait, wasn't as critical of intersex terminology as Vanessa, she still preferred DSD language: "I like *DSD* . . . because if you say *DSD*, [sex is] kind of camouflaged. When you say *intersex*, it has the word *sex* right in it and it's like, 'What?' It's kind of a red flag for people." When I asked Hannah if she felt gender was hard-wired or performative, she had a "both"/"and" explanation:

> I think it's a combination, I really do. . . . Because like sometimes I'm like, "I don't want to wear makeup," but I wear it to work because it's expected of me to wear it to work. But there are times where I'm in the mood and I'm like, "Oh, I wanna put makeup on," and everything. . . . But whenever I'm at work, and I'm like, "Okay, all these other women have makeup on, and I don't have makeup on." Because I will, sometimes, I'll be like, "Fuck it, I ain't wearing makeup," and I'll go two weeks without wearing a drop of makeup. And then finally I'll be like, "Ugh." I'll feel like I don't look as nice. Like they're all made up and pretty and here I am with my blotchy

skin from my HRT [hormone replacement therapy], you know what? My hair pulled back in a ponytail and I'll be like, "Ugh, I look gross, I don't feel pretty." So then I'll get with the program and start wearing makeup again [*laughing*].

When I asked Liz, a thirty-three-year-old woman with an intersex trait, how she felt about the new nomenclature, she replied, "I really don't care. I don't get involved in that kind of garbage." However, she never used the term *intersex* in her interview, wherein she also expressed a more essentialist understanding of gender: "Girls are more touchy-feely. Girls are more sensitive. Girls do think more internally. Guys kind of gloss over things. They are much more third party–type people. They like a lot of intensities. They're . . . um. . . . [G]irls are much more judgmental . . . I think." Jane, a fifty-four-year-old intersex woman well known in the community, also professed not to care, though she had a more strategic perspective. She did not understand why some intersex people so adamantly resisted DSD nomenclature, as she felt DSD terminology could provide a basis for productive conversations with medical professionals:

> I can be on the outside of the room arguing about terminology and if I embrace [DSD] and the door opens and let's have a real good substantive conversation because we are talking about the same thing . . . you can call me frog. I don't give a crap what you call me as long as we're moving forward advocating for families and advocating for small children that don't have a voice . . . so, when people want to argue 'til the cows come home that "*disorder is such an ugly word*" and "*we're not disorders . . . we're not disordered*" . . . oh, get the fuck over it.

Jane's primary concern was the treatment of intersex people, not the nature of gender. She explained:

> I think that a lot of sociologists and a lot of people who are involved in queer studies and gender studies are really locked into this mindset that gender is an evil social construct and we've got to tear down the binary, blah blah blah blah blah. And it irks me to no end when intersex conditions are used as fodder for some of those arguments. It really ticks my

ass . . . excuse my language. And like I said before, the world is populated primarily by men and women and we all find our place somewhere in that continuum. . . . And I just break out in hives when some magazine article comes out about that, or some writer wants to promote that, or some gender warrior intersex activist wants to promote that, or putting "I" on your driver's license and blah blah blah. I can't imagine anything more horrific than raising a child in our society as something other than a boy or a girl. Um, it's hard enough going to elementary school and high school and fitting in with your peer group. . . . Yes, gender is an issue but it's very secondary. And who gives a flying fuck where you fit in the continuum of male and female?

Well-known intersex activist Emi Koyama, founder of the national activist and advocacy organization Intersex Initiative, publicly endorsed DSD language on her blog for reasons similar to Jane's.[24] Emi explained: "In conclusion, I am endorsing the shift from 'intersex' to DSD not as a simple gesture of either defeat or confidence, but as a way to affect [sic] gradual reforms of the medical model that pathologizes intersexuality and simultaneously to collaborate, to build links with the tradition of radical disability activists and theorists who are seeking to uproot it."

Although many of my study's participants held strong views about DSD terminology, a minority was not committed to either *intersex* or *DSD*. Some believed that individuals should have the right to choose the terminology they prefer. They tended to see the use value of both intersex language and DSD nomenclature and usually avoided the debate altogether. Cheryl Chase supported others in their preferences: "I think people should use whatever term suits them. I think in a medical context, *intersex* is really counterproductive. It isn't a diagnosis. . . . It's totalizing, and the way in which it's totalizing causes doctors to be so freaked about it that they're going to lie. If that's the word that they get to use, they're not gonna use it, they're gonna lie about it. And we know that lies create shame." When I was diagnosed in the 1990s, medical providers were known for lying to patients about their intersex traits and encouraging parents to do the same. For example, doctors (and parents) would tell intersex people that they had been born with underdeveloped ovaries that were prone to cancer and thus needed to be surgically removed. One justification for this deception was that gender identity development would be disrupted if intersex people discovered they had an intersex trait. Chase argues that DSD nomenclature gives medical professionals a less politicized term for the diagnosis, which would in itself, she implies, possibly put an end to such lies. In contrast, Maria, a thirty-two-year-old intersex woman, had "mixed feelings," but "for technical reasons," she said, "I think *DSD* is appropriate. But as an activist, *intersex* really highlights . . . it really is different . . . it's just not some disorder." Mariela, a twenty-nine-year-old intersex woman who was introduced earlier, hadn't "put much thought" into terminology and preferred "either one, really. It's another label."

Individuals without strong terminological preferences did seem to adopt a more socially constructed understanding of gender. Emily, an intersex woman introduced earlier who was open to both intersex language and DSD terminology, said she used think of herself "as definitely more masculine," but that shifted as people started pointing out that she had "a lot of feminine qualities." When I asked her where these qualities came from, she said, "socialization . . . friends, family, watching the movies." Maria shared this socially constructed view of gender and extended it by critiquing the sex binary as an "oversimplification." When I asked Jenna, the thirty-one-year-old intersex woman introduced earlier, who also didn't have a strong language preference, whether she thought gender was hard-wired or performative, she explained: "I think it's probably a few different things. I think it's environmental . . . the roles you saw growing up, that's what you're going to tend to do. I don't know what kind of percentage to give it. Is it 40 percent environmental and 60 percent biological? I don't know. Maybe it's 30 percent environmental and 60 percent biological and 10 percent whatever you decide. You choose."

While *intersex* and *DSD* are both labels, it is important to note that intersex language is often used to express an identity that challenges the sex binary, while *DSD*, as medical terminology, upholds its existence. More often than not, those who preferred *intersex* identified as *lesbian, gay, bisexual, asexual,* or *queer*: 86 percent of the interview participants who identified as LGBA or Q preferred intersex terminology while the others either chose DSD language (9 percent) or were indifferent (5 percent). These individuals tended not to be concerned with fitting into a sex binary that mapped neatly onto gender and sexuality binaries; indeed, they typically rejected doing so. But while most were in what appeared to be same-gender relationships, this was not always the case. Several were living in one gender, usually as women, while partnered with a person on the other side of the gender spectrum, usually men. Leigh, who identified as queer, and Jeanne, who identified as bisexual, were both involved with men. Similarly, Chris, a man who identified as asexual, had been partnered with a woman before her untimely death. On the other hand, among the 32 percent of intersex people who identified as straight or heterosexual, 75 percent preferred DSD terminology. Given that DSD terminology rests on the idea that sex, gender, and sexuality correlate biologically, it makes sense that those who preferred it would choose a heterosexual identity and romantic relationships with people of the "opposite" gender identity.

Access to Biological Citizenship

Diagnostic nomenclature has to do with more than just words. In my data, those who embraced DSD nomenclature tended to have full access to biological citizenship, unlike those who preferred intersex language. Rose and Novas's (2005) conceptualization of biological citizenship at the individual level has to do with how individuals use biomedical language to describe aspects of the self—in our case, intersex traits. As Rose and Novas (2005) maintain, biological citizenship is established and controlled by powerful institutions, notably medicine, whose rules and expectations are adhered to by active biological citizens. These could include adopting a particular diet, maintaining a fitness routine to maximize health, or, in the case of the intersex community, adopting DSD nomenclature to describe intersex traits. Those who do not adhere to the rules and expectations of biological citizenship are viewed as problematic persons. Rose and Novas explain: "The enactment of such responsible behaviors has become routine and expected, built in to public health measures, producing new types of problematic persons—those who refuse to identify themselves with this responsible community of biological citizens" (2005, 451). In this section, I argue that the concept of biological citizenship can help us make sense of the fact that those who embraced DSD nomenclature tended to have positive relationships with medical professionals and family members, while for those who insisted on intersex language, those relationships were more likely to be fractured.

Intersex people who welcomed DSD terminology tended to report positive relationships with their medical providers. Liz explained that "[A doctor] cleared everything up. . . . I saw a couple of other doctors [in my city] that also cleared everything up." Vanessa said of her doctor: "She's really helpful, really helpful." Tara had similar experiences with medical professionals: "[My diagnosis] was straightforward. . . . The doctor was . . . really nice about it . . . supportive . . . summed it up like . . . you basically are born like a woman that had a hysterectomy . . . you just have to take estrogen to help with your bones . . . we're gonna remove your gonads."

Still, not all intersex people in support of DSD terminology felt comfortable seeking medical care. When I asked Hannah about her relationship with medical professionals, she explained:

> Today, it's all right. I will go to the doctor as a last resort. I don't like going to the doctor, which I'm sure [is shared by] a lot of people. . . . I think I have a little stronger phobia of doctors than your average Joe because the way I look at it is, I have very vivid memories of being in [a children's hospital], laying [sic] on a table with my legs spread open, and all these doctors and these interns or student residents or whatever. I remember them marching in through my room, looking at me down there, poking me, you know, like I was a sideshow freak, you know? There must have been at least twenty-something white coats com[ing] through there.

While Karen had less positive relationships with medical professionals at one time, she reported that this was no longer the case. "I've been treated like shit by doctors for a very long period of time [but] not currently." When I asked her what had changed, she said she started to tell doctors what she needed without being disrespectful: "I said, 'This is what I want. This is how I want to be handled. This is what I want you to do. I don't want you to stop being a doctor and not tell me the things I need to know, but I've been lied to in the past and I've been treated terribly and treated like a lab rat, and that's not going to happen here.' So he agreed and that's that."

Many of the intersex people who supported DSD terminology had never been involved in intersex activism, nor did they have any interest in constructing an intersex identity. Instead, they welcomed DSD language because it conceptualized intersexuality as a medical condition rather than as a social identity, a conceptualization that seems necessary for accessing biological citizenship. Because DSD nomenclature was formally introduced and thus defined by the medical profession, which along with other institutions (for example, the government) has the power to regulate and control biological citizenship, it makes sense that those who embraced DSD terminology tended to describe mostly positive relationships with medical professionals, with whom their self-understanding fell into line.

Access to biological citizenship might also account for the many stories of parental support shared by intersex people who embraced DSD terminology. Liz, who was not fond of intersex language, said that her mother "was supportive . . . she was just supportive . . . it was very good. She went through all the steps with me . . . took me to doctors and stuff. Very simple." Hannah had a similar experience with her parents, especially her father, with whom she spoke frequently when she first learned her diagnosis: "My dad was really there for me. He's the person that I turned to, having conversations, crying on the phone 'til 2 in the morning. While everybody else in the house was asleep, I was on the phone with him." Vanessa noted similarly that her parents "always have been supportive of [her] . . . and [tried] to make [her] a happy person." Like Liz, Vanessa did not particularly care for intersex terminology. Interestingly, Vanessa was careful to explain that soliciting the support she desired from her parents was an ongoing process: "I would say they're supportive. I think my dad is becoming increasingly so . . . I think my mom is still turning the other way and keeping her distance on this. She'll say, 'I'm here for you if you wanna talk,' but then if I wanna talk, she's sort of busy. . . . That's sort of what goes on with her."

Karen reported a generally positive relationship with her late mother, but their discussion of intersexuality was strained. Karen explained:

> I did talk to her about it and it was not a good conversation, although my mother and I always had a good relationship. She didn't remember a lot about it. . . . But I did make sure she understood that I had been lied to, and now the only part of it I couldn't quantify was how much she had lied to me. But I mean, [in] retrospect, my mother was very clearly outdone by these doctors. And I know that from having read pieces of my medical

record. Things like "We wanted to do such and such tests to [Karen], and [her mom] vehemently disagreed and absolutely forbid us to do it and it's just clear that she doesn't understand." . . . It was deprecating remarks like that about my mom.

Despite the difficulty of discussing her trait with her mother, Karen didn't appear to hold her mother responsible for the medical intervention performed on her. In fact, she seemed to stand up for her mother, criticizing how doctors had treated her.

Given that medical professionals promote DSD language, it makes sense that intersex people who reject that terminology tended to describe problematic relationships with them. To reject DSD language is to challenge the medical profession's biopower over one's body. Some people explicitly refused to defer to the prestige and authority that society grants medical professionals. As Chris put it, "I just find [medical experts] . . . dogmatic . . . they have it all figured out . . . and that doesn't sound like science to me. . . . Just because they have a 'doctor' in front of their name . . . when I was younger . . . I was a lot more respectful of that." Millarca stated simply, "I don't trust doctors." Caitlin described her relationship with medical professionals similarly:

> I don't like doctors. I don't go to the doctor very often. I don't trust doctors. That's a very triggering environment for me. I mean . . . I do have to tell every doctor I go to that I have [an intersex trait] because of course one of the first questions they ask you is "When was your last period?" So definitely in any medical setting where I'm a patient, that comes up immediately, and I really hate that. Like I had gone to the doctor because I broke my wrist, and I'm like, "I promise I'm not pregnant, I don't want to talk about it." And they're like, "But tell me why." And I'm like, "No, really, trust me, there's absolutely no way." . . . I don't necessarily want to always have to go into having [a discussion about my intersex trait] with every single doctor if I'm not there for anything related to that part of my body, you know? But I don't really trust doctors. Yeah, I don't know. I don't trust doctors as much as I should.

Although Irene also didn't trust doctors, she reported forming some positive relationships that were contingent on the doctors' willingness to relinquish at least some of their authority: "I have learned to pick doctors. I've had a lot of bad experiences, but in recent years . . . I had a woman doctor in [a geographical area] who was just amazing. . . . She's totally cool about everything and wants me to educate her and tell her as much as I can."

Such descriptions of troubled relationships with medical professionals formed a common theme among those who rejected DSD terminology. They challenged doctors' biopower, specifically the right to authoritatively define their bodies, as well as their positions as experts. Most of these individuals had been heavily involved in intersex activism in the 1990s, a period characterized by protests against medical professionals, and many were still involved in activism through appearances on television shows and in documentary films. All these elements made it unsurprising that they strongly opposed DSD terminology.

It was only slightly more surprising that they also reported fractured familial relationships. Parents often avoided the topic of intersexuality around their intersex children, which may have been one strain on their relationships. As fifty-seven-year-old Paul put it, "The parents know the elephant is in the room, and you're not really supposed to talk about the elephant in the room." He went on, "[I am] not in a very good relationship with my mom. . . . I think the whole thing is deeply troubling." Stevie also described a fractured familial relationship. She was estranged from her parents throughout her twenties and thirties because her parents attempted to police the formation of her gender identity: "My mother wanted to get me involved in social philanthropic things that would model . . . what a woman in society does. . . . [For example,] there was a modeling component to [an organization my mother got me involved in] . . . a "modellette program" [that] basically had beauty and poise training. . . . [My parents] knew they had a task to try to bring about a certain result . . . to raise this child as a girl." Stevie became reacquainted with her father only after her mother's death. Pidgeon, who claimed an intersex identity, explained that their parents attempted to influence their identity formation: "I think a lot of times, our parents are so scared that the doctors made the wrong decision and we're going to veer off to this other gender world . . . so they kind of police it. My parents didn't technically tell me all the time that 'You're a girl and you're going to be a girl,' but I'm sure it was always playing in the background of decision-making." The fractured familial relationships reported here were not likely caused by terminological preference but may have more to do with embracing, rather than rejecting, intersex as an identity component. For example, as shown above, it was not uncommon for intersex people to believe that their parents had attempted to influence their gender identity formations, which would have happened long before DSD language was in play. However, it seems logical to assume that the rejection of DSD terminology would only increase family conflict.

Involvement with intersex activism, past and/or current, might also help explain these fractured relationships. It's possible that these families view claiming an intersex identity and involvement in intersex activism as shameful public choices at odds with the biological citizenship they may have desired for their children, which depends upon accepting that intersex is a medical condition rather than a social identity. When I asked Millarca how her parents responded to her activism and public claim to her intersex identity, she replied:

> My family was ashamed. They thought that I shouldn't talk about things in the family outside of the family. So they didn't want to hear or watch the [intersex documentary I appeared in or the television program about intersex that I was on]. . . . [Today, my relationship with my family is] the same. It's still strange. We don't really associate very often . . . it's been like that most of my life. It's not like this just because I'm intersex, it's everything, like intersex, being gay, being into leather and S&M, and just not conforming to their politics.

While involvement in intersex activism may not be well received by family members, it can present an opportunity to discuss intersex with loved ones. Mercurio shared her experience:

> And when I did one of my shows, I came out to [my mother] and I say that sort of hesitantly because on one level she already knew but she was kind of choosing to not think about it. And I had to jog her memory about a lot of things, such as saying to me, "When you were born, they didn't know if you were a boy or a girl," which she did when I was about nine. And she wouldn't remember about these estrogen pills, which she fortunately argued with my father against me taking. He didn't [make me]

have the surgery done, but once I started menstruating, he immediately took me to a bunch of doctors and they told me I was going to have to take these pills every day to help me "grow." My mother later admitted that it was estrogen to make my breasts grow, blah blah blah. But, I actually had to do a show-and-tell with my mother because she was so insistent on saying, "Oh, you're just a normal girl!" And I was like, you know I don't feel abnormal, but my body is different.

Those who were noncommittal about DSD terminology did not display a clear pattern in their relationships with medical professionals. Fifty-year-old Emily explained that she was "distrusting" of medical professionals and took whatever they said "with a grain of salt." Skywalker had a similarly troubled relationship with doctors who encouraged her to "lose weight" and "let [her] hair grow out" in order to forge a feminine identity and attract men. But others had relationships with medical professionals that were as positive as those of individuals who embraced DSD language. Mariela, who did not claim an intersex identity but did not have strong feelings about terminology, described doctors as "very supportive . . . they did what they could, it was just me that didn't want to deal with it." Thirty-eight-year-old Kelly reported, "I have a very good relationship with my primary care physician."

Interestingly, those who were open to both terminologies tended to describe positive relationships with parents. For example, Skywalker "talk[s] to [her] mom a fair amount" about her trait, and her mother sends her supportive information gleaned from WebMD searches. Emily described her relationship with her parents as "good," adding that her parents are people she "get[s] along with." Jenna also has a positive relationship with her parents. She shared: "My parents were, like, as long as [I'm] happy, that's all that matters."

Participants who were indifferent to DSD terminology seemed to be capable of deflecting at least some of the impact of the biopower embedded in the diagnostic nomenclature. By not exclusively embracing either label, they were free to use whichever term they found more effective in a given setting, which gave them the same access to biological citizenship as those who exclusively embraced DSD language, though that access may have been tinged with more distrust of the medical establishment. They could easily shift to the DSD terminology that constructs intersex as a medical problem in order to fulfill their medical and relational needs. At the same time, they could use the power of intersex language to debunk the logic of the sex binary if they began to feel abnormal. However, although the choice to straddle the line between intersex language and DSD terminology is not inherently problematic, it does raise the liability of being challenged by people who are passionate about either terminology. There are, for example, people who refuse to engage with the AIS-DSD Support Group because *DSD* is part of the organization's new name. At the other end of the spectrum, people who are anti–intersex terminology have asked me to stop using what they consider offensive language. These challenges make it difficult to maintain a commitment to linguistic flexibility.

Conclusion

Intersex people are born with bodies that directly challenge the sex binary. While this can offer a space for liberation, it can also be a very challenging place to live, work, and play. Not surprisingly, all of the intersex people I interviewed reported that they had struggled with their bodies at some point in their lives. Many of these struggles were tied to the emotional and physical consequences of having undergone irreversible surgical interventions under the guise of medical necessity. However, as I've shown, even those who escaped surgery reported struggles. Cultural critic and longtime intersex advocate Iain Morland (2011) offers some insight into why this might be: "[A] remedy for genital ambiguity, if one were needed, might be gender certainty. After all, if ambiguity is social, and if gender is a social construction that includes expectations about anatomy, then being certain about one's gender might dissolve anxieties about one's genitalia, entirely without surgery taking place" (152). If we accept Morland's assumption and work toward reducing intersex struggles by making it known that gender certainty is independent of bodies, I think we would be on the right track but would still fall short if we didn't simultaneously engage with the idea that diagnoses are also socially constructed.

I contend that relying solely on pathologizing medical terminology to describe our bodies also contributes significantly to the struggles intersex people face. History has shown that there are implications to naming and defining traits as disorders.[25] Consider, for example, attention deficit hyperactivity disorder (ADHD). Peter Conrad (2007) argued that the expansion of the ADHD diagnosis to include adults who hadn't previously been diagnosed had lasting implications for how those individuals understood and explained their behaviors. In his classic work on asylums, Erving Goffman (1959) eloquently shows how individuals who were labeled as mentally ill were forced to adopt identities consistent with their diagnoses. When diagnosis and identity merge, there is little room for reconsidering the diagnosis, which makes it difficult for an individual who is diagnosed with a mental disorder to escape the label, even when the symptoms that originally gave rise to the diagnosis disappear.[26] Individuals diagnosed with post-traumatic stress disorder or medically labeled as alcoholics face similar situations.[27] In all of these cases, the label imposes an identity that is believed to coincide with the diagnosis, whether or not the diagnosis is justified.[28]

Every intersex person I spoke with was familiar with DSD nomenclature, which underscores how quickly it has become ubiquitous. Yet DSD language has not been uniformly accepted by the individuals whose bodies it is meant to describe. Because the medical profession—a powerful institutional authority—formally introduced the DSD nomenclature, people must engage with it, but the way they engage can either allow or prevent access to biological citizenship. In order to be active biological citizens, intersex people are expected to adhere to the nomenclature, but this poses a significant problem, because its very wording, *disorder of sex development*, clashes, for many, with their conceptualizations of gender and senses of self as healthy people.

Those who choose intersex language, rather than DSD terminology, typically conceptualize gender as a social construction that we all perform. This understanding seems to put them at ease in their bodies, but it comes with a cost: lack of access to biological citizenship and the relational support from medical providers and family members it entails. The individuals in my study who embraced DSD terminology tended to have a more essentialist understanding of gender. They also struggled with their gendered selves, worrying that they weren't legitimately gendered, which led to anxiety about their bodies and, in intimate settings, their genitalia. On the positive side, these individuals seemed able to access biological citizenship, which tended to afford positive relational support from medical providers and family members. These supportive relationships gave them comfort in knowing that they were not alone as they navigated their diagnoses, yet it did not necessarily mitigate their anxieties.

For people with intersex traits, DSD nomenclature is the key to accessing biological citizenship and the right to relational support it makes possible. This alone ought to be enough evidence that there is indeed *power in a name*. My concern here has little to do with whether biological citizenship is good or bad for the intersex community; rather, I am bothered that in order to access biological citizenship, intersex people must openly accept and adopt a linguistic formulation—*disorder of sex development*—that requires them to see themselves as living with an abnormality.

The crucial exception to this rule was those who felt people should be able to choose whichever term they desired, a flexible strategy that clearly has use value. Those who straddle the terminological options, rather than exclusively adopt one label over the other, can selectively employ the language they believe will be most beneficial in any particular situation: They can embrace intersex language to thwart negative self-understandings and use DSD nomenclature to facilitate positive relationships with parents and medical professionals, thus achieving a more open access to biological citizenship. However, in order to achieve this open access, one must acknowledge that gender is a socially constructed system of stratification[29] *and* that a medical *condition* is only as real as its definition.[30]

Given these findings, I suggest that, rather than allow the power embedded within medical nomenclature to further divide our community, we need to set aside our terminological differences and come together to reduce the shame, stigma, and secrecy associated with our bodies, a goal that we all share. The only way we can accomplish this, however, is by collectively understanding both gender and medical diagnoses as socially constructed phenomena. If we can convey this complex understanding to ourselves and the society at large, we can blunt the medical profession's exercise of institutional power over our bodies.

5

A Different Kind of Information

When I started this project, I didn't know how having a child with an intersex trait emotionally affects parents. I soon learned that one of their main feelings is guilt, which emerges from a number of factors, ranging from the belief that they are genetically responsible for the intersex trait, to regret that they allowed medical providers to perform irreversible surgeries on their children's bodies. Alexis, the parent of a young intersex adult, said, "One of my problems today is that I have guilt because they did surgeries and stuff on her. . . . We never talked about it and that was so wrong of me. I never brought it up as a parent, but I was just . . . leave it alone. She's not saying anything. I don't want to upset her. Why open up? I feel guilty." Hope, whose teenager has an intersex trait, echoed Alexis: "I also feel guilty that I gave her this disorder. Even though it really wasn't my fault, I just feel guilty. So that's the truth. I'm just saying the truth."

Speaking with parents like Alexis and Hope made me wonder how my parents experienced intersex. My parents knew about my intersex trait long before I did, but they followed doctors' orders and chose to keep it a secret, never discussing it. In *Making Sense of Intersex*, philosopher Ellen Feder (2014) reports that the parents she interviewed "shared an interest in maintaining the secret of intersex because [they feared] others' ignorance or cruelty could harm their children. . . . The silence that parents have maintained to protect their children protects parents, too, from a kind of guilt by association" (62). Did my parents keep my intersex trait a secret in order to protect us? Did they ever wonder if my intersex trait was somehow their fault? Did my parents feel guilty for consenting to my medically unnecessary surgery? Was their own guilt the reason we never discussed my diagnosis or medical history? In August 2011, at the age of thirty, I sat down with my parents and asked them these questions for the very first time. One reason we had to have this conversation then was that they wanted to attend my dissertation defense. My dissertation was about intersex, and I was concerned that they would feel uncomfortable moving directly from years and years of private silence about my intersex trait to listening to me discuss my research on the intersex community in a room full of sociology faculty and graduate students. My trepidation was only heightened by the fact that I planned to open my dissertation defense by discussing how my personal experience with intersex had fueled my professional interest. My parents, it turned out, were determined to attend my dissertation defense at any cost, even if they had to confront their own feelings about my intersex trait. While my family continues to avoid discussions about intersex, this project has allowed me to understand more deeply how intersex affects the entire family,[1] which in turn has helped me to understand my parents, who are not unique in the intersex community.

In this chapter, I focus on parents at the interactional level of gender structure.[2] The interactional level of gender structure is where relationships and expectations concerning gender are formed. It's also where individuals reinforce or challenge the gender structure, with assistance or resistance from others. I am specifically interested in how medical providers interact with the parents of intersex children and, in turn, how these interactions shape the medical decisions parents make on behalf of their minor children. I address several questions in this chapter: How do medical professionals present the intersex diagnosis to parents of newly diagnosed children? What happens when parents receive information directly from the intersex community? Does such information alter their view of medical professionals and their recommendations? How do parents feel about DSD nomenclature, and do their views on terminology have anything to do with their understanding of gender?

These questions are informed by several theoretical frames, including Nikolas Rose and Carlos Novas's (2005) theorization of biocitizenship with regard to activism (rights biocitizenship), the Internet (digital biocitizenship), and specialized information about one's body (informational biocitizenship), discussed at length in previous chapters, and Giorgio Agamben's (2005, 2000) discussion of state of exception, which I encountered through earlier work with my colleague Erin Murphy (see Davis and Murphy 2013). Agamben articulates the state of exception as a physical place in which a sovereign suspends legal rights because they deem it necessary. In terms of intersex, the state of exception is the intersex body. As explained in chapter 3, medical professionals tend to hold binary and essentialist ideologies about sex, gender, and sexuality. Medical authority at the institutional level of gender structure deploys these ideologies to allow doctors to save the intersex body by medically "fixing" the intersex trait. A key piece of this process occurs at the interactional level of gender structure when doctors present intersex as an abnormal medical emergency that necessitates immediate attention, in particular to parents of children born with intersex traits. Constructing intersex as a medical emergency allows doctors to circumvent professional medical ethics that would normally prevent them from performing swift, irreversible, and medically unnecessary surgeries on children's bodies.[3]

Psychologist Vickie Pasterski and colleagues (2014) recently concluded that "both mothers and fathers [of intersex children] reported overall levels of PTSS [post-traumatic stress symptoms] that were comparable to those reported by parents of children diagnosed with other disorders, in this case cancer" (373). My interviews reveal that parents feel pressured into consenting to medical procedures, despite their lack of complete information, such as knowledge about intersex shared by other families and intersex adults. Later, they feel decisional regret.[4] This insight can help us make sense of Pasterski et al.'s findings. More specifically, in this chapter I show that parents desire peer support in the form of more information about intersex—that is, not exclusively from medical providers but from other parents and even intersex adults. This leads me to argue for the necessity of ensuring that parents receive information directly from the intersex community. Parents who are connected to the intersex community make better-informed health care decisions for their children, which, in the long term, can minimize or eliminate decisional regret. Parents who have access to such information are also better positioned to challenge medical authority at the interactional level of gender structure by questioning medical recommendations and/or their urgency. Parents with decisional regret are also positioned to challenge medical authority, but they tend to do so by becoming advocates who advise other parents to proceed cautiously with medical recommendations.

Although almost all of the parents I interviewed were skeptical of medical authority in one way or another, they nevertheless expressed support for DSD terminology. As we saw in chapter 4, many intersex adults resist this terminology, because they believe its language of *disorder* pathologizes the body. Parents, however, are less concerned with the possible pathologization of their children's bodies and more interested in distinguishing intersex from lesbian, gay, bisexual, and transgender issues.

Treating Intersex and Evading Responsibility

As soon as medical professionals suspect that a patient has an intersex trait—whether they notice ambiguous external genitalia at birth or discover a trait like complete androgen insensitivity syndrome (which leaves people with an external female appearance despite internal testes and XY chromosomes) during adolescence—they tend to turn the situation into an emergency. When I asked Dr. D. to describe what happens when she encounters a baby with an intersex trait, she explained, "We try to find out as much biochemical and genetic data as we can, *as fast as we can*. We look at the phenotypic appearance of the exterior of the child. We try to figure out . . . if we know their biochemical basis or what we think it is, what is likely to happen to them at puberty" (emphasis added). If the hospital has a DSD team in place, it usually convenes to discuss medical care and treatment plans. Dr. I. describes a typical meeting: "[We] meet as a team and think about what are the options, which option we feel is medically in [the child's] best interest, and then we present the options to the family. And then we help the family reach a decision as soon as possible." Ellen Feder (2014) names this process nondirective counseling and describes it as "[involving] the provision of information a health-care provider believes to be important for weighing various possible interventions in a given medical situation and what is known of the outcomes without directing the person counseled to make a particular decision" (134). However, as Feder notes, this style of care, especially when providers present intersex as a problem, might explain why parents would consent to medically unnecessary and irreversible procedures, as surgical interventions allegedly can "fix" the "abnormality." Parents who chose to leave their child's body as is would be, in essence, "choosing abnormality" but only because intersex was presented by providers as an abnormality (Feder 2014, 152). Given such, I see providers as decision makers and not experts enacting nondirective counseling. Furthermore, as my interview with Dr. I. revealed, providers meet as a team to discuss "which option [they] feel is medically in [the child's] best interest." In other words, when providers meet with the family, they've already reached a decision. How is this nondirective counseling if providers present intersex to parents as a problem, meet as a team without parents present to decide on the best available option, and then, and only then, ask parents to make a life-altering decision for their child? It seems to me that this nondirective counseling approach is really a move by providers to, consciously or not, evade responsibility for their interventions.

If providers truly wish to engage in nondirective counseling, parents should be allowed in DSD team meetings. I have seen doctors express significant uncertainty about treatment options during and after DSD team meetings, without parents there to witness it. If parents were included in DSD team meetings, they would be exposed to medical uncertainty, they would experience first-hand the arbitrary nature of many medical recommendations, and they might be inclined to slow down their own desire for immediate medical response to their child's intersex trait. Including parents in this preliminary decision-making process would also address the pressure doctors claim parents place on them to quickly "fix" their children (discussed below). Instead, when the DSD team approaches parents with a recommendation after resolving their medical uncertainty, doctors are able to mask that uncertainty and dangerously evade responsibility and oversimplify the decision-making process by suggesting that they are simply offering options in the form of nondirective counseling.

I also find it troubling that DSD team meetings commonly include photographs of children's genitals. Dr. F. explained to me that providers are "looking at the genital structures . . . [we] see photos, not the actual child." This practice is problematic in two ways. First, when doctors consult with one another over photographs of patients' genitals, rather than interact with actual intersex people or their parents, they objectify their patients, which makes it easier for them to remove themselves emotionally from the situation. While this may make it easier for them to resolve medical uncertainty, it also divorces them from the emotional—and even the physical—realities of the people whose lives they seek to improve. Second, although photographs may be used to reduce the number of genital examinations, photography itself can be quite traumatic for intersex people, as my ethnographic observations revealed. At one continuing medical education event, which included intersex adults, parents, and providers, a number of intersex adults criticized a doctor for showing photographs of her patients' genitalia in her research presentation. One intersex woman sitting near me gasped and shook her head, then whispered to the person sitting next to her, "Can you believe this?" During the question-and-answer period, this same woman told the doctor that she was very uncomfortable throughout the presentation because of the photographs, which she said were "triggering," given her own experience with medical photography. The 2006 "Consensus Statement on Management of Intersex Disorders" acknowledges this problem, cautiously stating that while "[m]edical photography has its place for record keeping and education . . . [r]epeated examination of the genitalia, including medical photography, may be experienced as deeply shaming" (Lee et al. 2006, 492–93). Nevertheless, DSD team meetings and medical research presentations still use photography.

As doctors urgently meet behind the scenes, privately exercising what sociologist Phil Brown (1990) describes as diagnostic technique (classification of the condition) that leads to diagnostic work (medical tasks to address the condition), parents naturally worry that something is seriously wrong with their child. Dr. I. elaborated:

The family is aware that [the DSD team is] getting additional data. We have to wait for labs to come back, [details about chromosomes] to come back. We let the family know that the emergency, which would have been a salt-wasting CAH, is or is not the concern. [Congenital adrenal hyperplasia is grouped into either salt-wasting CAH or simple virilizing CAH, with the former posing a greater health risk as a result of too much sodium being lost in the urine.] Once you say there's no medical emergency here, then we say, let's get some more data. And then we get more data. . . . We [then] meet as a team and think about what are the options, which option we feel is medically in [the child's] best interest, and then we present the options to the family. And then we help the family reach a decision as soon as possible.

Salt-wasting CAH requires immediate medical attention, but once it is ruled out, the emergency situation that has been constructed does not end. Instead, as Dr. I. explains, the search for "more data" continues, reinforcing the assumption that intersex is a medical emergency, even though intersex traits rarely need immediate medical attention.[5] Furthermore, as sociologist Steve Epstein (1996) cautions in *Impure Science: AIDS, Activism, and the Politics of Knowledge*, when medical providers conduct data searches, patients tend to become experimental subjects rather than benefit from the data collected.

Providers seem largely unaware of the role of their medical authority in this process, framing their efforts as simply presenting information that enables parents to make decisions (i.e., nondirective counseling) and ignoring the ways in which they limit and/or influence those decisions. According to Dr. B., "We kind of go through, here are the two choices, and get parents pretty actively involved. I think sometimes we overwhelm them a little bit with information, but I don't know how else you do it, I'm not sure." Dr. I. further explained, "So for family, it's hard. They are forced to make certain decisions— not all decisions—some decisions with the best available data we have at this time. And that's hard. That's what parents do all the time. We just do the best we can [in our recommendations] with the data we have." When I asked Dr. B. what could be done to make the decision-making process easier for parents, she suggested "better" data: "I mean, I'm really saying that we could give better information to parents 'cause they make the decision." Dr. A. was convinced that all doctors do is share information:

> I generally think it's up to parents to pick [a sex and ultimately a gender] based upon what information we have available at the time. The most important counseling that we can give to people is [to] help them understand that the gender of rearing may or may not turn out to be much related to sexual preference and behavior later on. And parents need to appreciate the ambiguity involved. . . . Our obligation as healthcare providers is to provide people with very complete information about what we

know about the biology of DSDs, about the implications of DSDs for later development of sexual preference and sexual identity.

Yet this "very complete information" is usually exclusively medical, omitting other valuable information described later in this chapter. This is hardly nondirective counseling. Although some medical professionals, like Dr. I., recognize that making decisions about medical intervention is difficult for parents, they still tend to place responsibility for those decisions entirely on the parents, excluding their own role in the process and, indeed, their responsibility.[6]

In fact, many doctors justify performing irreversible surgeries by claiming that they act exclusively on parental wishes to align a child's sex and gender. Dr. G. described one instance: "The hard part [for this family] is that every time this mom changes the diaper of her baby girl, she sees these testes . . . it's this daily reminder. Some families could accept that, but for this family, it's just really getting debilitating." Dr. A. elaborates: "[S]ome families, for cultural, religious, or psychological reasons, may feel very strongly about the importance of trying to have their child look more typically male or female. Under those circumstances, I would counsel them to defer surgery; I wouldn't oppose surgery." Interestingly, the parental concern with aligning the intersex child's assigned sex and gender, described by Dr. G., Dr. A., and others, continues even as children age. Dr. D. noted that "parents [of intersex children raised as boys] complain to me that they wanna wear their sisters' dresses and play with the dolls; they don't wanna go out and play with other boys."

It may be true that parents initially want their child's sex and gender to be aligned, but providers can and should do more to help parents understand that sex, gender, and sexuality are not neatly correlated characteristics. It is, of course, not surprising that they don't, given that so many of them buy into the same narrow ideologies that influence parental desires, believing that sex, gender, and sexuality are biologically prescribed in and on our bodies (see chapter 3). However, these beliefs should not excuse medical professionals from being culturally competent providers. If, for example, it were common practice for doctors to explain to parents that sex, gender, and sexuality are not biologically correlated (i.e., genitalia is not predictive of gender performance or sexuality) and are in fact quite variable characteristics in themselves (e.g., penises and vaginas are not all identical, meanings attached to gender presentation vary, and sexuality can be fluid throughout one's life), parents might understand and respond to their child's intersex trait in different ways.[7] This is not an entirely new recommendation. Social psychologist Brendan Gough and colleagues argued in 2008 that medical providers who treat intersex need a more complicated view of sex and gender: "A more fluid understanding of sex and gender would perhaps help parents cope with the initial impact of having an intersex baby. . . . dominant cultural definitions of sex and gender need to be challenged, and health professionals should be trained to accommodate manifold permutations of sex so that parents of intersex children can be better supported" (504).

Although doctors tend to claim that they are merely information providers, it is important to keep in mind that they make treatment recommendations from a position of power and authority over the intersex "emergency" they create. This leaves parents inclined to accept medical recommendations and simultaneously allows providers to evade responsibility for their actions. Caroline Sanders and colleagues (2012) exemplify this in a recent publication that, on the one hand, encourages nurses to familiarize themselves with the debate concerning intersex surgeries but, on the other hand, counters this progressive step by, in an all-too-familiar fashion, suggesting that parents are responsible when intersex individuals come to regret the procedures performed on their bodies. They state that "being [parents] does not necessarily mean that they will always make good decisions for their child; their wishes may be in conflict with what their child as a young person or adult might have chosen for themselves" (Sanders et al. 2012, 3320).

Leaving medical professionals out of this account of the decision-making process places responsibility fully on parents. In reality, however, medical providers play a critical role in the parental decision process, and they need to own this. In a fascinating experimental study, biomedical ethicist Dr. Jürg Streuli and colleagues (2013) enlisted eighty-nine third-year medical students to play the role of "parents" of intersex children. They were randomly assigned to watch either a medicalized video in which an "endocrinologist" discussed intersex or a demedicalized video in which a "psychologist" discussed intersex. After watching the video, they completed a questionnaire that assessed their understanding of intersex and asked them to decide whether "their" child should have surgery. The subjects who watched the medicalized video were significantly more likely to opt for surgery than those who watched the demedicalized video. These findings illustrate the critical role providers play in the parental decision-making process.

Attempts to assess whether parents are comfortable with the surgical decisions they've made on behalf of their intersex children largely report satisfaction (see, e.g., Crissman et al. 2011; Dayner et al. 2004). For example, in an article that appeared in *The Journal of Urology*, Dr. Jennifer Dayner and colleagues (2004) report that parents are willing to grant consent to surgery at any cost, even loss of sexual pleasure: "When parents whose children had undergone genital surgery were asked to rank sexual responsiveness and genital appearance by indicating whether they would have consented to genital surgery even if a reduction in sexual responsiveness were certain, most (95%) indicated that they would consent to surgery" (Dayner et al. 2004, 1764). However, in direct opposition to these findings, my interviews with parents revealed that parents who consented to medically unnecessary interventions tended to express decisional regret. Alexis shared: "The part that kills me today is the surgery. We've met a lot of people in the community that didn't have any surgery done with their kids and, yeah, you know, I blamed myself . . . for many years, many years. We should have left [our child] alone. Parents should go out there early and get educated, very educated. Look into everything and find out everything. Stay on top of it." I wish I could say that Alexis was an outlier in my study, but that would simply not be true, as many of the parents I interviewed, especially those whose children were of adolescent age or older, expressed similar decisional regret. How could my interviews with parents yield different findings than those presented by medical providers? It can't be that my parents were mostly recruited from peer support groups, as the same is true for the participants in Dayner et al.'s study. It seems to me that our different research findings might have to do with the fact that I interviewed parents who had intersex children of all ages, from young babies to grown adults. On the other hand, Dayner et al.'s study included only parents whose intersex children were between the ages of eight months and thirteen years old. Maybe if Dayner et al. had surveyed parents who had older intersex children, who would arguably be more informed and better equipped to question why their parents consented to such procedures, their findings would be more in line with mine.

Many of the parents I spoke with who reported feeling guilty for allowing doctors to operate on their intersex children also felt that their decisions were anything but informed. Susan asserted that "parents aren't making an informed decision" when doctors treat it "like oh you got to do it right now." Michelle shared a similar experience: "They told us they needed to remove the gonads at one month. So she had surgery because they said 'This is what you do' and we had no clue. Now I wish that we [had] waited just because of all the information that we have now." Jen, a parent of an intersex teenager, explained, "The only reason we did [the surgery] then was[,] that was what we were told eleven years ago. [The doctor said:] 'Oh you have [the testes] taken out because . . . you either make a decision, or you wait for [your] teen to decide'. . . . [The] word *cancer* came up enough and that did it. So we took them out." Alexis was also scared into consent. She said that she agreed to surgery because the doctors assured her it would emotionally benefit her child: "They told me the surgery would fix [her intersex trait]. Everything was going to be fine. . . . They kind of, like, forced surgery. They would say, 'When she gets older, it's going to look funny.' You know that I didn't want her to have, you know, it looked like a penis." Yet even though Alexis acknowledged that she felt the doctors "forced surgery," she still, years later, blamed herself for granting consent, emotionally declaring, "We should have never had it done, down in the vaginal area."

Parents of intersex children face an important decision—to consent or refuse to consent to irreversible medical interventions—with, in most cases, no personal experience with intersex to guide their decision. Many parents have never even heard of intersex prior to their children's diagnoses. They thus need as much information as possible before consenting to procedures, or else they may eventually experience guilt and decisional regret. But parents do not need more medical information—they have enough of that. Instead, as I show in the next section, they need a different kind of information, which comes from peer support.

A Different Kind of Information

The intersex community offers parents valuable peer support in the form of empathy and reassurance as well as, and perhaps just as valuably, information.[8] When they meet members of the community, parents hear personal accounts of intersex from other parents and intersex adults. Intersex adults often tell them that the surgeries they endured as children were harmful, rather than helpful. Other parents give them advice about navigating medical recommendations. These personal experiences provide parents with a different kind of information than the type they get from doctors.

One of the first things parents discover when they connect with the intersex community is that most intersex adults in the community are openly critical of unnecessary medical interventions. Stevie, a forty-four-year-old intersex woman, shared with me that the surgical procedures performed on her body when she was a baby created an incredible amount of trauma in her life. She explained, "The damage is sort of done. . . . So in the process of fixing me, [doctors] broke me." Given such trauma, Millarca, a forty-six-year-old intersex woman, had this advice for parents:

> If [we are talking about a] child, I would tell the parents to get educated, for one. For two, make an informed decision regarding surgery, meaning that if it's not medically necessary, I wouldn't allow my child's clitoris to be shaved or cut off. . . . I wouldn't listen to a white heterosexual doctor that tells you that your child's clitoris should be cut off . . . because [your child's sexual stimulation is] not their concern. Their concern is not the child's sexual stimulation because she's going to be married female anyway and her sexual stimulation doesn't matter; it's her husband's pleasure that matters. [Parents have] to be smart enough to make an informed decision. [They] should talk to as many people as possible. So be informed, make a proper decision, and talk to as many people as possible.

Parents offer similar advice about delaying or refusing medically unnecessary surgical interventions. Hope said she would advise other parents to "Just breathe. Take a deep breath and don't feel like you have to rush into anything. . . . Let [your child be]. Get all the information you can and then wait. Don't feel like you have to rush and make your decisions."

There are many benefits beyond the decision-making process to connecting with other parents of intersex children. Many of the parents I spoke with described the incredible emotional value of knowing other parents of intersex children.[9] Most of them met through support groups, specifically the Androgen Insensitivity Syndrome Support Group–USA (AISSG-USA), described at length in chapter 2, or *dsd*families.[10] Today, a simple Google search can lead families to intersex organizations all over the world that allow parents of newly diagnosed children to connect to the community and learn about their children's intersex traits and the broader experience of intersex directly from others who are also personally affected by intersex.[11]

AISSG-USA, parents explained, was an important factor in learning about and coming to understand intersex. Laura, a parent, said "My big turnaround was when I met you guys [in the support group]. It showed me that it all works out in the end. It does all work out. You guys all lead great lives and you're not so worried all the time." Drew had a similar experience: "I think the [intersex support group] conference reassured me the most. To see . . . I mean, those girls, it seems like they're all accomplished people that are doctors, people that are whatever they want to be. Nothing's going to hold them back." Other parents echoed Laura's and Drew's relief in knowing that many intersex adults live happy, healthy, and productive lives. Susan elaborated, "Meeting successful [intersex] adults is like knowing that you have a kid who can grow up to be whatever they want."

Feeling that they weren't alone was another positive aspect of the support group experience, for both parents and their children. George, a parent of a young adult with an intersex trait, explained: "[The support group] eased our fears. It allowed us to meet people who have similar conditions or who have gone through similar things as our family and as our daughter, and it just shows that we're just part of many people, which is great. [We've since] been involved with the [support group] board to help plan for things in the future." Alexis said, "Attending the support group was a great experience. It was so nice to meet other parents and girls . . . you know, going through the same thing. I think it was really helpful. . . . I think it was good for us to hear that [my child] wasn't the only one." Jeff had a similarly positive view: "It's a very extremely comfortable environment and it's almost therapeutic, basically. . . . There's been so many helpful things. I keep the different e-mails [from other parents]. I save them in different files, based on what they're on. For example, surgery, whether they have to have any surgery or not. There has been a lot [of e-mail communication] on hormone replacement therapy.

Stuff like that, and so it's good to be able to reach out and hear those different parents talking sometimes." Jen, the parent of a teenager, felt the support group benefited both her and her daughter: "It's actually enriched my life, both of our lives. Being a part of this group is very enriching and very empowering. And it feels so wonderful to be part of such an amazing group of people."

Given their experiences, it is not surprising that when I asked the parents I interviewed what advice they would give to parents of newly diagnosed children, they highly recommended connecting with the intersex community. Michelle passionately recommended joining a support group because of the comfort she felt in connecting with others who understood first-hand what it meant to have a child with an intersex trait: "Definitely sign up and be a part of the support group. You need to find others who understand, and that's the support group." Sue added: "Come out to one of [the annual support group] conferences! Talk to someone who's been there and done that! . . . If you don't show up, you don't know what you're going to be missing. And if you don't show up, then other people may really not benefit from getting to know you. I mean, not getting to know *us*, but we may not benefit from getting to know somebody like *you*." Hope would also encourage parents to get connected: "Get on the [AISSG-USA] support group website. And get the information [about your child's trait] from there because that's probably the most helpful information. They have so much!" Susan found solace in the support group, but she also acknowledged the importance of medical experts, albeit without necessarily deferring to their expertise: "I'd say, talk to the experts. Of course, a lot of doctors aren't experts, but their opinion could be helpful. Talk to other parents. That's my big recommendation. Talk to the adults, too. What would they do differently? Parents can benefit from informing themselves, educating themselves, and just being patient. But, yes, definitely talk to other parents. I'd strongly encourage parents to get advice from other parents about what to say to their child about their condition, how to say it, and when to say it."

This universal enthusiasm was expressed even by parents who had only recently encountered the community. I met Marty, a middle-aged mother whose teenage daughter had been recently diagnosed with an intersex trait, at her very first AISSG-USA meeting. She found the support group by "doing some research on the web" even before her daughter was officially diagnosed. Shortly after we met, on the very first day of the meeting, she agreed to be interviewed for my project. Before we knew it, we were sitting in the hotel lobby, candidly discussing the emotionally exhausting but liberating weekend we had just experienced. As in all my parent interviews, I asked Marty what advice she would offer other parents of newly diagnosed children. Without hesitation, Marty replied, "Definitely come to a meeting, a support group. And get your children involved in the support group meeting. And . . . I guess obtain your child's medical records, I think that would be good." Sheryl, another parent who had only recently connected to the support group, explained that the group left her, her husband, and her daughter who'd been born with an intersex trait feeling that they had "friends for life." Her husband, Michael, added, "I think initially . . . it was just this feeling that wow there is this group of people like our daughter. It is filled with supporters who have this common interest to be part of a group and not feel so outside, and also the group has greater sources of information about [our daughter's] condition than most doctors."

Research has demonstrated the benefits of support groups for intersex people. Dr. Peter Lee and Dr. Christopher Houk (2012, 2010), two well-known medical experts on intersex, acknowledge the importance of "social support" in the "satisfactory levels of social and sexual function" that intersex adults express (Lee and Houk 2010, 1).[12] There is no doubt that parents also experience benefits. Psychologist Matthew Malouf and Dr. Arlene Baratz, Family and Medical Adviser to the AIS-DSD Support Group, had this to say about support groups: "Support groups are often excellent resources for families, adolescents, or even adults looking for information about DSD. Support groups offer a degree of anonymity and privacy (through registration requirements) and can help individuals find referrals, answer questions about insurance coverage and care, and connect with others in varied geographic locations" (2012, 84). In a special intersex issue of *Psychology & Sexuality*, Baratz collaborated with others (Anne Tamar-Mattis, Katharine Baratz Dalke, and Katrina Karkazis) and added:

> In our experience, quality peer support is very effective for helping parents through both the early days and the long-term challenges of raising a

child with a DSD. Many families have been grateful for peer support, and clinicians should direct parents to peer support groups (monitored by experienced parent educators and expert clinicians), where they can learn about the lived experience of families. High-quality peer and psychosocial support can prove invaluable for helping parents to make treatment decisions by clarifying the ways in which DSD and its treatment may impact a child's friendships, school experiences, adolescent development and intimate romantic and sexual relationships. (Tamar-Mattis et al. 2014, 49–50)

Unfortunately, many providers do not refer their patients to support groups for any number of reasons, whether they are primary providers who aren't experts on intersex and thus are unaware of the groups' existence, or they fear that peer groups will taint parents' views about medical providers and their recommendations.

All of the parents quoted above are connected to the intersex community, most of them through the AISSG-USA. This raises the question of parents who, for whatever reasons, are not connected to the intersex community. The only parents I interviewed who were not connected to a support group were ex-partners Sarah and Steve, introduced in chapter 1, whom I met through their daughter, who is connected with the intersex community. Neither was familiar with the support group, and they did not even know such groups existed. This is not surprising, given the reluctance of many medical experts to refer parents (or their intersex children) to support groups, but combined with the fact that Sarah and Steve didn't have access to the Internet or know how to use it to find information, it is also a reminder of the potential socioeconomic limitations of my research conclusions.

It is worth noting again that almost everyone I spoke to for this project was economically positioned to have access to health care. Except for Sarah and Steve, many also had the benefit of a college education and the cultural capital it embodies, including access to the tools to seek out advocacy and support organizations. But Internet access and skill aren't the only barriers to one's benefiting from the intersex community. Membership in the AISSG-USA and attendance at its annual conference, to give other examples, are each quite costly. Laura, the parent of a young child with an intersex trait, expressed empathy for those less privileged than she. When I asked how her experience might have been different if she and her husband weren't college-educated and worked on a factory line, she answered, "It would be very different. I know one particular parent who struggles. We correspond on the e-mail list, but she's never been able to attend a meeting. I think it's hard for her. It's hard to help her. I really want to, I want to try to offer her whatever help [I can]." Laura's husband, John, elaborated, "And what kind of stinks about those folks, back to the person on the line, making for example $10.35 an hour, it's difficult for them to spend a $50 registration fee, $50 membership fee, $250 fee for the conference, $100 a night for the hotel, so . . ." Laura jumped in to add, "It's a strain. For us, it's a strain. I can't imagine what it's like for others. But I think it is important. [Interviewer: "I know they have scholarships . . ."] But, if everyone applies, they'll all be gone."

A substantial amount of skill (Internet searches, for example) and resources (access to the Internet, money to travel to a conference, etc.) is needed to find and connect with the intersex community. If medical providers don't connect parents with peer support and parents aren't equipped to discover such resources on their own, they may never benefit from the community. Those who do find themselves in touch with the intersex community but don't have the resources to access it fully will not gain the full benefit of things like in-person support group meetings. AISSG-USA offers scholarships to cover the expenses of attending the annual conference, but only a limited number are available and every year people get turned away because there are far more applicants than there are scholarship funds available.[13]

Another interesting feature of intersex support groups is that they bring together people from various cultural backgrounds. While this is generally considered to be a benefit, it can be challenging at times. Shelby, the parent of a young child with an intersex trait, explained:

> One of the hard things about it is that all of these parents are different. This is the only thing we have in common with one another. So there are some people [in the group] who are very religious, and are very . . . . They respond to questions posed by other parents saying "Well, you have to leave it in God's hands. This was the card that you were dealt." And I'm just not comfortable with that. I'm like, that's not what I believe, that's not what I think because I think that if it was in God's hands, everybody

would be perfect, you know? I don't honestly believe that God makes people imperfect for a reason. . . . I'm not down with that. So I struggle with that, and there's a lot of that [in the support group]. I think that there are a lot of folks that are very "Jesus made her this way" and I am like, no, scientifically that doesn't jive with me. . . . And I've now finally taken a few steps back.

Still, despite issues of access and different belief systems, there is no question that parents can benefit significantly from access to peer support through the intersex community.

When Parents Resist Medical Authority

When parents of newly diagnosed children hear personal accounts from other parents and intersex adults, they gain the information they need to think critically about medical recommendations that are grounded in the fallacy that intersex poses a medical emergency and necessitates immediate medical intervention. Laura and John bring unique insights to bear on this experience because they were pressured into an immediate decision about their daughter's care and then, after meeting with the support group, became comfortable with reconsidering future medical recommendations. Not long before we met, their daughter had exploratory surgery to repair what was assumed to be a hernia. During the procedure, their daughter's surgeon came to the waiting room. Laura told me that "you could just see it written on his face . . . he wouldn't tell us what the problem was." John jumped in: "Yeah, he just said, 'It's not right, it's not developed correctly . . . We need to take it out and have it tested.'" Laura interrupted, "He said, 'It's more testicular' . . . but won't tell us what he thinks it is." What happened next illustrates how parents make immediate decisions when intersex is framed as a medical emergency:

> LAURA: So he said, "But your daughter doesn't need it, so we're going to take it out. You want me to take it out, right?" And so we had to make that decision to take it out when we don't know *anything* about it. . . . She's under anesthesia and we got to go "yes" or "no." "Yes" or "no."
>
> JOHN: Anesthesia and cut open.

LAURA: He said, "She doesn't need it, it looks more testicular, you have
a daughter and it looks more testicular. What are you going to do?"
JOHN: It would have been nice at that point if he came out and said,
"Look . . . we're going to do a biopsy of it, put it back in, sew her up,
and she may need another surgery later."
LAURA: He didn't even give us that option. It was really a peer pressure
type thing. . . . And then . . . ugh . . . the worst part was being in the
room afterwards, the recovery room we were in. All of the nurses
were coming in and nobody would talk to us. They wouldn't look at
us in any way. The doctor came in to check her and wouldn't look at
us. And then they would be constantly pulling her out to make sure
she looked like a female, and checking the incision site. But they
wouldn't look at us. . . . They were checking her vagina out.

The doctor removed only their daughter's protruding testis during
that initial surgery. This left the question of what to do about the other
testis. Laura and John were initially in conflict, but after meeting with
the intersex community they came to agreement. Laura explained:

John wanted [anything male] gone. Anything male, he wanted it out of
our daughter 'cause he considered it male, and our daughter a separate
entity. . . . But the more I was reading, and I guess on the [intersex support group] webpage and then on the parent group talking about it. . . .
"Your daughter's natural hormone maker [should be left alone] . . . it's
better than being on a pill all your life. Sometimes kids with the pill have
a difficult or more difficult puberty." There wasn't any doubt about it that
I wanted to keep [her as is after that]. . . . [I]t took me a little while to
convince John we needed to keep it.

John added, "I'm glad Laura did all the research that she did and found
the support group and knocked me across the head [*laughing*]."
Like Laura and John, Shelby and her husband, Drew, parents of a
baby with an intersex trait, were influenced by personal accounts from
the intersex community. After hearing from an intersex adult, they refused immediate surgical intervention. Drew explained, "When we met
[an intersex adult], he said . . . 'Hey, keep [the testes] in as long as you
can.' I think I'm going to listen to a guy like that a little bit more. . . . [I]t

sounds like there's probably more benefits to having them than [fewer] benefits." Drew and Shelby experienced some pushback from medical experts for resisting their recommendations. One doctor told them that their daughter "would be very psychologically damaged if [they] don't remove her testes immediately." Jeff, a father of a young adult with an intersex trait who hasn't had surgery, had a similar experience. He described with obvious emotion the advice of medical professionals:

> Basically immediate surgery, for certain things . . . and you know we found out that that wasn't really necessary. . . . They didn't get into whether [she] would need surgery or not need surgery. . . . The big fear right away is . . . this is the fear that the pediatric endocrinologist put into us, is "Oh she's got to have it taken out right now or she's going to have cancer. Right now." And then when you start finding out [from] people who've dealt with these situations for a long time, over many, many years . . . they're like "That's not true." The risk is no different than [for] anybody else.[14]

Perspectives from the intersex community are important in these situations because they help parents get the information they need to delay or outright refuse medically unnecessary interventions.

Sheryl and Michael, parents of a teenager with an intersex trait, agreed to irreversible surgical interventions when the intersex trait was presented as a cancer risk. Sheryl recalled:

> [The doctors] had their ideas . . . their theories. . . . [We were told that the gonads] were a possible cancer threat. Okay, fine. So there was not much question about [going ahead with the surgery with that information]. . . . [Now,] I'm kinda like, I don't think they know everything and I think there are definitely ego issues that bug me. . . . These doctors, they want you to think that they're experts in the field. They're not. The support group has been a *huge* resource. . . . [When we saw the doctors in the major city that our suburban doctor referred us to] for the first time [and told them about the support group], they were very kind of like, "Don't believe everything you hear at those meetings." And they were a little like, "You don't need to [get involved with the support group]". . . . When we said, "Yeah, we're going to their meeting," they said, "Don't believe everything you hear at those meetings."

As noted above, many medical providers view support groups negatively, perhaps because the information shared at meetings can lead parents to challenge medical recommendations.

Indeed, empowered with knowledge from the intersex community, many of the parents I interviewed started questioning medical professionals. Laura explained:

> After I went to [the support group meeting], I said, "Am I right for feeling this way? Am I right for feeling that [the doctor] is not taking care of us like he should?" At least just be respectful, you know? . . . I actually took information to give to our endocrinologist at our Children's Hospital in the state—which is pretty well known. I actually took a copy of the [DSD handbook] to give to them for their future patients, because they didn't have that. They didn't have a pamphlet. They didn't have *anything.*

John, Laura's husband, added, "It would've been nice if they . . . could've just said, 'Look here's the [support group] websites.'" After attending a support group meeting, Jeff also questioned his doctor's ability to handle his daughter's care: "We decided that our doctor—obviously I'm not going to criticize him that much—but they didn't have any experience in children with AIS, nor did I feel that they really wanted to do the research to find anything out, which was the most frustrating part, because I did the research and I'm not the doctor. They could've done the research. Why should I have to do it for them?"

Although Susan was also critical of medical professionals, she was a bit more understanding: "It just depends on how much the doctors know, how old they are, and their philosophy. People go through medical school at different times. . . . What I'd like to see is that basically people are completely comfortable with it. Doctors across the board are current, and everybody is in agreement about how to manage medical issues. There's no bias. People are just comfortable. People realize that everybody's different. It doesn't matter whether it's your eye color or your hair color or whether it's your sex." Susan provides an important reminder that medical providers are not a monolithic group of individuals. Their views may vary by age, though my interview sample does not provide evidence of this, because most of the providers I interviewed were middle-aged or older and had long since graduated from medical school. This raises the important question of how, if at all, intersex medical care will change if new medical school graduates—especially those born in the 1980s or later, who may be more open to diverse understandings of sex, gender, and sexuality—specialize in intersex.

Words, Words, Words

The parents I interviewed have embraced the intersex community, with some even using the information they gained there to challenge medical authority and question medical recommendations. I thus wondered how they would feel about the terminological shift from *intersex* to *disorder of sex development*. Given that these parents tend to be critical of the medical profession, I anticipated at least some resistance to DSD terminology. Jeff directly expressed the resistance I expected:

> I prefer the *intersex* term because the other term, *disorders of sex develop-ment* . . . to me . . . suggests a negative. Is it really a disorder? I don't know if it really it is. It's . . . it's just normal genetics. That's the way I see it. . . . Sometimes people have four fingers. I mean, it's not . . . to me . . . it's not a disorder necessarily. I've got a daughter that has CAIS [complete androgen insensitivity syndrome], okay? I know there's a lot of other worse situations, okay, and maybe for that reason that term is a better fit for them. So I guess I'm on . . . if there are two sides of the bookends, I'm probably more on the *intersex* one just because I feel that it's less . . . negative of a term.

But Jeff's criticism of DSD terminology as "negative" was not uniformly shared by other parents in the intersex community. Drew explained:

> *Disorder of sex development* is safe to say. It is what it is. The condition hasn't changed. You're trying to get the outside world outside of our community to understand it better so you want to use different terms that make other people feel comfortable. That's the way I look at it. *Hermaph-rodite* . . . I don't know. It seems like a bad word. I never really thought about it too much. I don't know why it seems like a bad word. I guess you associate it with people that were a little bit "out there," not necessarily [part of] mainstream society.

Shelby, Drew's wife, elaborated:

> There was all of this stigma surrounding *hermaphrodite, testicular femi-*
> *nization*, like all these terms that are often used to describe what's wrong.
> When people would get aghast at hearing those terms, they'd be like "Oh,
> they're outdated" and it's just like, "No, they're just a term." And that's
> how you have to think of it. It's just a term. Yes, she does have testicles, so
> *testicular feminization* makes sense to me as a term. You know? Calling
> it this 'complete androgen insensitivity' . . . like that to me too is all these
> words that they are using to describe it, but let's just say what it is 'cause
> that's what it is. She's a female that has testes. I don't see why there should
> be this stigma I guess surrounding those words. That's all they are, words
> [*laughing*].

As Shelby continued, it became clear that the terminology itself didn't
bother her. Rather, what she found troubling was the way in which *I*
for *intersex* was being included in the LGBT abbreviation, resulting in
LGBTI.

> The stuff that bothered me in the beginning [with the terminology] was
> the parallels to the gay community. And that still kind of bothers me a
> little bit. I don't have a problem with people who are homosexual, I don't.
> But, I would not consider [my daughter] to be in the same community. I
> don't like that she's clumped into that. It bugs me that she's clumped into
> that same community. . . . I saw a lot of *intersex* [language] in doing my
> research and it kind of bugs me a little. There was this movement to have
> *M*, *F*, and *I* on forms for people so that they can check off if they're inter-
> sex, and you know that would be great for people who are transsexuals.
> I was really—excuse my French—pissed off in reading that because I'm
> like, "No, gender is what's in here most of the time too and how you feel
> about yourself." It's not necessarily what you look like. *Intersex* shouldn't
> even be a term to me.

While Shelby states she doesn't have a problem with homosexuals, she
is still keen to distinguish her daughter from homosexuals and trans
people, a distinction that she fears the *I* attached to the LGBT abbreviation would erase.

Sheryl and Michael also expressed some reluctance to clump *intersex* together with LGBT:

MICHAEL: I haven't thought about [terminology] before, but I think I would like *DSD*. I prefer it. Why? It seems more like a medical diagnosis. You say something like *intersex* and the person goes "What?" and they're thinking . . .

SHERYL: Tranny?

MICHAEL: It's such a . . . I don't know.

SHERYL: But then it's a token disorder. See, I spend a lot of time with people with disabilities and I always feel like whatever you want, that's what we'll use. Whatever *you* pick, that's what we'll use.

MICHAEL: But it depends on the context that you have to use those terms. If you're talking to your classmates, if you're talking to a doctor . . . it depends on the context.

SHERYL: I think [our daughter] says *intersex*, but I'd have to ask her.

Many parents prefer DSD terminology because they don't want their children to be perceived as trans. *Disorder of sex development* implies that their child has a medical condition, which may not be how they see trans.

Other parents did not directly tie their dissatisfaction with the term *intersex* to the LGBT abbreviation. In fact, they often found it hard to articulate the reason for their preference. George and Sue described their gut discomfort:

SUE: Personally, the *intersex* term . . . like the first time we heard it was a real . . .

GEORGE: It had a lot of baggage.

SUE: It was kind of a deer-in-the-headlights term to me. . . . They talked about intersex and that was a little "Ahhhh!" It made my hair stand up 'cause I was there with my little girl and I'm going like, "Oh my gosh!"

Michelle, a parent of a teenager, added: "I like *DSD* . . . *disorders of sex development* . . . and I don't know why I prefer that. *Intersex* is like . . . what's that? So it's definitely *DSD* that I prefer." Marty also expressed preference for DSD terminology over intersex language, and while she didn't mention homosexuality, she did note that *intersex* triggered sexual associations: "Yeah, I like *DSD*. I don't like that *intersex* word. I don't know why. It just sounds like intercourse and sex [*laughing*]. I think that's part of it."[15] Laura had a similar view: "I don't care for [*intersex*] a whole lot. It's like a confusion word, it's like a sex confusion, and that's not what it is." John, Laura's husband, added: "Out in the public, *DSD* would be fine because not everybody knows what that is . . ."

Other parents had no strong preference for either intersex language or DSD terminology. Susan had no interest in the issue: "What seems silly to me is that some of this is just language . . . like what do you call it? *Intersex* or *DSD*? It's like, let's not get hung up on labels." Some parents, like Jen, ultimately felt it was not up to them:

> [My daughter] has always said, "I have AIS," the condition. So *intersex* or *DSD* . . . we've never really adopted either one of those. We've always said *AIS* or *complete AIS*. That's what we've always chosen to use. And we've always called it a condition, not a disorder, because some people believe you can fix a disorder. So yeah, that's what we've always used. I mean whatever the affected person wants to have their condition termed as . . . well, that's up to them. You know? They shouldn't be classified as one thing or another.

This lack of terminological consensus among parents in some ways mirrors the preferences of intersex adults. In chapter 4, I discussed how some intersex adults I interviewed rejected DSD language, others embraced it, and a few were indifferent to it, and I suggested that people should be able to use whatever terms they prefer.

But while the nomenclature preferences of intersex adults could, for the most part, be mapped onto their understandings of gender, this was not the case for the parents I interviewed. Much to my surprise, parents in the intersex community tended to hold a more socially constructed view of gender. In other words, they recognized the plasticity of gender. For example, Marty, who preferred DSD terminology, told me that her daughter's diagnosis was changing how she conceptualized gender:

> It's just more confusing. It's not black or white. I don't know. It's a whole new way of thinking . . . because, to me, it's not about activities. I think

men can mother and women can do the roughest, toughest activities. . . . One of the biggest things I've come away with is this whole continuum thing, like maybe you're a woman but you have masculine characteristics, or maybe you're a man with feminine characteristics. It's not black or white, it's much more gray. That's new to me. I've never felt that way, especially coming from a religious background. You kind of think that it's this or it's that. There's no gray.

Susan, who believed people should use whichever term they desire, also expressed a more socially constructed view of gender, though she believed biology also had something to do with it: "I think part of it is how you raise them. But I think part of it is innate. Boys tend to be drawn towards playing with masculine-type toys . . . trucks, planes, or whatever. Girls tend to be drawn towards playing with dolls, playing house, and being the mother. Part of that may be, 'I'm a girl, this is what my mother does, so therefore I should be more like that.' But part of it I think is just innate." Even as these parents expressed socially constructed views of gender by recognizing the plasticity of gender, they tended to reject intersex language by criticizing its addition to the LGBT abbreviation.

This pattern of parents' expressing more expansive views of gender is both similar to and different from the observations made by sociologist Tey Meadow (2011), who studied the parents of transgender and gender variant children. Meadow concludes: "Biomedicine, psychology and secular spirituality offer tools for constructing a normative view of gender, and yet individuals can employ these restrictive knowledge systems in the service of expanding gender ideologies, not constricting them" (2011, 742). On the one hand, it seems that many parents of intersex children in my study are similar to those in Meadow's study because they biomedically frame their children's intersex trait as a *disorder of sex development* but in turn expand their understanding of gender norms. If parents of intersex children desire that their children be viewed as "normal" kids, they need to expand their understanding of gender in such a way that rejects gender essentialist logic—which assumes that sex and gender, and arguably even sexuality, are strictly correlated. In other words, if parents are raising their child who has complete androgen insensitivity syndrome as a girl even though she was born with internal testes and XY chromosomes, they need to disentangle sex from gender, otherwise their daughter with testes and XY chromosomes will be viewed as abnormal. In the process of disentangling sex from gender, parents expand gender ideologies by embracing a more socially constructed view of gender. On the other hand, some of the parents of intersex children I interviewed expressed homophobia and transphobia, unlike the parents of transgender and gender variant children in Meadow's study. Parents of intersex children may be using biomedical discourse to normalize their kids, but this process seems only to expand rather than disrupt gender norms. Parents want to make it clear that *intersex* is not included in the LGBT abbreviation, a concern that does not, in the same way or to the same level, seem to affect the terminological preferences of intersex adults.[16]

Conclusion

In a medical culture wherein irreversible and medically unnecessary procedures are still being performed on the bodies of intersex children, it is important to present parents with the most complete information possible—including the fact that they can choose to do nothing. Providers need to make clear that a child with an intersex trait that is not life-threatening is born perfectly healthy, without the need for any type of intervention, especially immediately. When doctors present the birth of an intersex baby as a medical emergency, they construct intersex bodies as what Agamben (2005, 2000) describes as states of exception. This framing allows providers to justify recommending and implementing medically unnecessary procedures, and it may cause parents, as Feder (2014) argues, to choose surgery in order to "fix" their child's "abnormality." Doctors who do this aren't acting out of a desire to cause harm. Rather, as I explained in chapter 3, their essentialist understandings of sex, gender, and sexuality lead them to approach the intersex trait as an abnormality that keeps individuals from comfortably fitting into these presumably natural and correlated binary phenomena. This allows them to justify the irreversible surgeries they recommend to parents and ultimately perform on the bodies of healthy babies and children. As social psychologist Brendan Gough and colleagues (2008) noted years ago, medical providers need a more complex understanding of sex and gender (and sexuality) in order to "facilitate better service provision and, ultimately, [strive for] greater informed consent" (493). Instead, as it stands today, by framing intersex as an emergency, rather than a normal variation, medical professionals establish a situation that only they, as authorities, are fit to respond to. In turn, parents tend to defer to medical expertise and, for the most part, blindly follow medical recommendations that lead them to consent to medically unnecessary and irreversible surgeries aimed at "normalizing" the intersex body.

Medical professionals rarely accept responsibility for the decisions they force onto parents. Instead, doctors tend to inundate parents with *medical information*, most notably myths about cancer risks associated with intersex traits. But what I have argued in this chapter is that parents of intersex children do not need more medical information. Instead, as I have shown, they benefit most from a *different kind of information* that emerges from the personal experiences of those in the intersex community—especially intersex adults, but also other parents. When parents have access to information that originates in the intersex community rather than in the medical profession, they are able to exercise what Rose and Novas (2005) call informational biocitizenship, which empowers them to make autonomous decisions about the bodies of the children for whom they are responsible—decisions that more often respect the autonomy of those children. This learning process often happens in the biosocial community of intersex support groups, support group websites, and affiliated e-mail lists, which means that we can think of informational biocitizenship as made possible by digital biocitizenship, which Rose and Novas describe as the electronic linking of individuals with shared experiences. Together, digital and informational biocitizenship enables parents to delay or even refuse medical recommendations, which usually improves the physical health of children and the mental health of both parents and children.

But my research shows that even as parents align themselves with the intersex community with regard to medical intervention, they remain more open than many intersex adults to the medical profession's DSD terminology. In fact, many parents found comfort in the DSD nomenclature and preferred it to intersex language, although for many this preference seemed related to the fear that their child might be associated with LGBT communities. It is also the case, however, that parents do find comfort in information that originates in laboratories and radiology rooms. They want to know, and rightfully so, that their children are healthy. But there are many different aspects to health, and to keep their children healthy over the long term, they also need the kind of information that can come only from the intersex community. Consider this plea to medical professionals that Ellie Magritte, a parent of a child with an intersex trait, shared in the *Journal of Pediatric Urology*:

> Caring for a child with genital difference means you are always on alert, but once you get used to it, and once your child gets older, it is not so hard. What is difficult is having to do it amid the isolation that can come all too easily from our children's diagnosis, and this is where we parents need your help, as medical professionals, in stimulating contact between affected families, in helping us share these experiences, with each other, with you and with the many parents who are too afraid to reach out to each other. (2012, 574)[17]

The intersex community needs medical allies. We need doctors to refer their patients for peer support in order for us to help one another realize that intersex isn't a medical emergency but rather a difference that ought to be celebrated. I also genuinely believe that providers need us, the members of the intersex community—especially if they wish to provide the best possible care for their patients and their patients' families.

6

Conclusion

The Dubious Diagnosis

I started *Contesting Intersex* with one broad question in mind: How did *intersex* become a *disorder of sex development*? I've argued that a good part of the answer to this question has to do with medical authority and jurisdiction over the intersex body. While Cheryl Chase and her allies came up with and advocated for the new DSD terminology, hoping it would improve intersex medical care, the phrase was formally introduced by the medical profession. DSD nomenclature initiated the transformation of intersex advocacy from collective confrontation to contested collaboration. Today, DSD terminology has come to replace intersex language in almost all corners of the medical profession, a change that dismays at least some members of the intersex community.

The rapid and widespread acceptance of *DSD* by medical providers has been striking, but it can be explained by its context. As I explained in previous chapters, when the terminological shift occurred in 2005, medical control over intersex was in jeopardy as a result of a confluence of factors, including feminist critiques of intersex medical care, the exposure of medical pioneer John Money as a research fraud, and intersex activism. As public awareness rose about the prevalence of medically unnecessary surgeries on intersex people, it seemed impossible that medical providers could continue to treat intersex traits as they had been doing. But when intersex traits were reinvented as disorders of sex development—itself an act of powerful evidence for the social construction of medical diagnoses—medical providers had the perfect opportunity to reassert their authority over intersex and reclaim jurisdiction over the intersex body in treating this new disorder. The fact that many medical providers hold essentialist understandings of sex, gender, and sexuality—tending to see all three as biologically prescribed and, in many cases, neatly correlated—made them more open to and quick to accept the idea of *disorders* of sex development, not to mention the pathologization its language implies.

But at the same time as DSD nomenclature has taken off in the medical profession, it has been a major source of tension in the intersex community, especially since it was officially introduced by the medical profession in the "Consensus Statement on Management of Intersex Disorders" (Houk et al. 2006; Hughes et al. 2006; Lee et al. 2006). Some intersex people accept DSD language, which allows them access to biological citizenship and its benefits, including supportive relationships with medical providers and family members, though at the cost of feeling stigmatized for being labeled as having a disorder.[1] Intersex people who reject DSD language tend to reject the idea that sex, gender, and sexuality are biologically prescribed bodily phenomena and instead embrace intersex itself as a powerful identity. In doing so, they challenge gender essentialist ideologies. However, I argue that by rejecting DSD terminology, they also limit their access to biological citizenship and its benefits.

Though DSD nomenclature has heightened divides in the intersex community, most intersex people agree that surgeries designed to "normalize" the intersex body shouldn't be performed until individuals are old enough to make their own decisions. This is an important goal our community shares, grounded in a desire to improve the lives of future generations of intersex people. But we need to remember that surgical interventions are not the exclusive cause of our struggles as intersex people, for even those who have escaped surgery struggle with intersex. This leads me to conclude that those struggles are not solely the result of dangerous interactions with the scalpel. Rather, they emerge out of the broader medicalization process, specifically the ongoing struggle to challenge the belief that we are abnormal, a belief held, alas, by too many of us, as well as much of the medical profession and society at large.

Given this challenge, our intersex community cannot afford in-group conflict over terminology. We need to be conscious of the origins and history of DSD terminology, but when we think about what needs to change to benefit future generations of intersex people, people should be able to choose whatever term—or terms—they find suitable: *intersex*; *intersex traits*; *intersex conditions*; *intersexed*; *intersexual*; *disorder of sex development*; *difference of sex development*; the abbreviation *DSD*; *diverse reproductive development (DRD); diverse sex development,* also known as *intersex (DSDI); intersex,* also known as *differences in reproductive development (IDSD); conditions affecting reproductive development (CARD); congenital atypical reproductive development (CARD); conditions related to reproductive development (CRD); variations of reproductive development (VRD); hermaphrodite;* a term we have not yet come up with; or no term at all.[2] This flexibility will provide us with more open access to biological citizenship, which I have shown has significant, though not exclusive, value. It will allow us to devote our time to other matters, such as ending the shame and secrecy tied to the intersex diagnosis.

Like some of my research participants, I find the word *disorder* pathologizing, but if strategically employing DSD nomenclature results in higher-quality medical care, why shouldn't we use it or some variation of it, including *differences* of sex development?[3] While I find value in Audre Lorde's (1984) claim that the "master's tools will never dismantle the master's house" (112), meaning that using the tools of oppression can't end oppression, sometimes we need to access powerful institutions in order to create change. Because *intersex* has historically been situated in the medical institution, we need to access that institution, which means that sometimes we need to work with its terms, even *disorder of sex development.* But we can control how we approach that term by acknowledging—and pushing medical professionals to acknowledge— that sex, gender, and sexuality, as well as medical diagnoses, are socially constructed phenomena embedded within a system of stratification. There is power in a name, and the intersex community can and should use that power to its advantage.

Intersex in the Spotlight

Since the publication of the "Consensus Statement on Management of Intersex Disorders" (Houk et al. 2006; Hughes et al. 2006; Lee et al. 2006), the global intersex community has experienced several significant events that have raised public awareness about intersex in ways that have significant potential to shift the definition, status, and experience of intersex for years to come. The years 2012 and 2013, in particular, brought new political visibility to the intersex community. In 2012, the Swiss National Advisory Commission on Biomedical Ethics[4] and the

German Ethics Council[5] independently offered opinions that critiqued the contemporary state of intersex medical care. In 2013, the United Nations Special Rapporteur on Torture condemned the medical profession's nonconsensual surgical treatment of intersex by associating it with torture.[6] The New Jersey Senate approved a bill that will allow intersex and trans people to change the gender listed on their birth certificates without surgery,[7] and a new German law came into effect that allows parents of intersex children to register their children as "X," rather than "M" or "F," on birth certificates.[8] Members of the intersex community often disagree on the implications of these events—for instance, the new German law has been criticized[9] for forcing intersex children into an arbitrary third option ("X"), unlike the law passed in Australia, also in 2013, that allows adults to choose a third gender designation on government-issued personal documents.[10] Still, there is no question that intersex has been in the spotlight.

In the media arena, in 2014 MTV's dramedy *Faking It* put intersex front and center by revealing that Lauren, one of its main characters, is intersex. As he developed Lauren's storyline, Carter Covington, the show's creator, worked closely with Advocates for Informed Choice (AIC), a legal advocacy organization quickly establishing itself as a global leader in the fight for intersex rights. Covington explained his motivation in an interview with *The Hollywood Reporter*:

> Part of the overall theme of *Faking It* is how hard it is to be your authentic self and how important it is to strive to do that. At the beginning of last season, we were discussing in the writers [*sic*] room what that could be and we stumbled on, "What if she were born intersex?" . . . We felt like it was a no brainer because it really frames who Lauren is and why she has walls up, why she is hyper-feminine and why she is who she is. It's a story I've never heard told before, and our show is all about showing the diversity of experiences that young people are faced with today.[11]

Lauren's big reveal occurred during the season 2 premiere. During the days leading up to and after the premiere, Inter/Act, AIC's new program for intersex people between the ages of fourteen and twenty-five, was extremely active in raising intersex awareness online. On their Tumblr page, they shared an image of Lauren with the Inter/Act logo (see Figure 6.1).

Figure 6.1. Lauren from MTV's *Faking It*. Photo circulated by Advocates for Informed Choice's Inter/Act members.

When the show was trending on Twitter and people started asking questions about intersex, Inter/Act members were prepared with an intersex FAQ (see Figure 6.2).[12] The FAQ addressed what intersex is, its estimated frequency, and the like. Inter/Act also provided definitions of key terms, including *ambiguous genitalia*, *informed consent*, and others. In their discussion of terminology, they did not use the word *disorder*, presumably strategically:

Intersex and DSD (difference of sex development) are the two current terms that most people use interchangeably. However, they both are controversial for different people. Some of our youth feel more comfortable with DSD as it might be the only term they are familiar with, while others prefer intersex over DSD. All intersex folks have the right to self define themselves at any particular point in their journey. It's better for people to come to their own conclusions about how they want to identify, rather than be told or pushed into identifying a certain way. If you don't know how someone identifies, feel free to ask![13]

With more than one million viewers during the season 2 premiere, according to the Nielsen ratings, the show undeniably increased the visibility of the intersex community.[14] Presumably this impact will continue, as Kimberly Zieselman, AIC's executive director, remains a consultant to the show.

Another 2014 television event, this one in the "news" realm, called attention to the intersex community, though initially in a negative way. On February 14, 2014, *Fox & Friends* co-hosts Elisabeth Hasselbeck, Clayton Morris, and Tucker Carlson "expressed incredulity in their on-air banter" as they discussed Facebook's addition of more than fifty gender options.[15] In response to this mockery of the intersex community, Sean Saifa Wall, AIC's board president, and I called out the co-hosts in an op-ed for *Advocate.com*: "Chances are many of the employees of Fox News know someone in their personal lives who was born with an intersex trait. At the very least, with this letter, we are assured that they know at least two people born with an intersex trait who are regularly made to feel abnormal because of their unique bodies. Before laughing at folks like us and the intersex community to which we belong, we encourage Fox News to exercise caution over this very serious issue."[16] Inter/Act also released their own op-ed at GLAAD:

What your anchors discussed on air actually highlights an important point. Not all people with intersex conditions identify as being "between" or "other than" male or female: many people consider themselves as men or women with DSD or intersex medical conditions. Many times, people misconceive what "intersex" or DSD means and—like your anchors—feel uncomfortable discussing it. . . . We felt disregarded as a butt of a joke. We felt when watching the clip that not only did your team not know what intersex was, but they also didn't find it worthwhile to learn more. People with intersex conditions face challenges every day. These challenges

Figure 6.2. Inter/Act's intersex FAQ, featuring three of its members. Photo circulated by Advocates for Informed Choice's Inter/Act members.

could happen when required to choose between "Sex—male or female" on a form. It could happen when sharing our intersex condition with a friend and being met with a blank stare. It happens every time intersex conditions are misunderstood and fetishized, in print and on television. These problems can only be solved in a permanent way when we increase awareness of intersex conditions and the issues we face as young people with these conditions and bodies.[17]

Several days later, and much to our surprise, *Fox & Friends'* Clayton Morris offered an on-air public apology: "Look, I made a pretty ignorant statement a few weeks ago . . . we were talking about the Facebook story . . . where they added a bunch of gender identifying things . . . and I made sort of an off-handed comment . . . and I regretted it later . . . 'cause wait a second there are people who are actually dealing with this . . . and I'm an idiot for saying something stupid like that, so before you open your mouth think about it a little bit."[18] That a co-host of a Fox television show would respond publicly—and apologetically!—to a critique from intersex activists suggests that intersex is indeed entering a new public era.

Intersex has also received attention in the sports world. In May 2011 and June 2012, respectively, the International Association of Athletics Federations (IAAF) and the International Olympic Committee (IOC) developed similar eligibility policies for women athletes who have hyperandrogenism, which they define as a higher-than-"normal" testosterone level, a "condition" often associated with intersex people who have not had surgical or hormonal procedures to "fix" their intersex trait.[19] In short, these policies prohibit women athletes with hyperandrogenism from competing with other women, unless they undergo invasive and medically unnecessary interventions, such as surgery to remove testes, to lower their natural testosterone levels.

These polices were established in response to the case of Caster Semenya, the South African runner who, after winning the 800-meter race at the 2009 Berlin World Championships, was banned from competition when sports officials discovered she had an intersex trait that they believed gave her an unfair advantage over other women athletes.[20] Semenya was eventually allowed to return to competition, but only after a media frenzy that took an emotional toll on the young athlete.[21] Colleagues and I have argued that there is no scientific basis for these policies, which derive instead from illogical thinking about fairness in women's sports (Karkazis et al. 2012). In particular, we point out that the policies are based on problematic assumptions about testosterone in women's bodies, including acceptable natural levels and effect on athletic performance.

Since the IAAF and IOC policies were implemented, four young athletes, all from developing countries, have been subjected to medically unnecessary interventions in order to compete (see Jordan-Young et al. 2014). In July 2014, the policies struck their latest victim, Dutee Chand, an eighteen-year-old Indian sprinter, who was targeted because of her natural testosterone level. However, Chand has filed a formal appeal in the Court of Arbitration for Sport, specifically challenging the policies on the grounds that they are "highly unscientific and unethical."[22] A "Let Dutee Run" campaign in support of Chand, including a Change.org petition, was initiated by three advocates: Payoshni Mitra, a gender and sports expert appointed by the Sports Authority of India to advise Chand; former Olympian Bruce Kidd, a professor of kinesiology at the University of Toronto; and Katrina Karkazis, a medical anthropologist and bioethicist.[23] Besides the hope that Chand's challenge will affect the policies, these events are drawing valuable attention to the nature and complexities of intersex.

But perhaps the most significant event of recent years took place not in a legislature, the media, or the sports world, but in a courtroom.[24] In 2013, Pamela and John Mark Crawford filed a lawsuit, in both federal and state courts, against "South Carolina Department of Social Services (SCDSS), Greenville Hospital System, Medical University of South Carolina and individual employees," on behalf of their adopted eight-year-old son, M.C., who was born with an intersex trait.[25] Before the Crawfords adopted M.C., he was in the South Carolina foster care system, where medical providers, with the support of social service employees, performed surgery on him at the age of sixteen months to address his intersex trait. According to the lawsuit, the surgery removed "healthy genital tissue" with the result of "potentially sterilizing him and greatly reducing if not eliminating his sexual function."[26] Several lawyers have come together to represent the Crawfords in this lawsuit, including Anne Tamar-Mattis (the Legal Director at AIC), the Southern Poverty Law Center, and two law firms working pro bono. The complaint filed in federal court states:

> This lawsuit challenges the decision by government officials and doctors to perform an irreversible, painful, and medically unnecessary sex assignment surgery on a sixteen-month-old child in state custody. Defendants performed this surgery for the purpose of "assigning" the child the female gender despite their own conclusion that he "was a true hermaphrodite but that there was no compelling reason that she should either be male or female." . . . Since a young age, M.C. has shown strong signs of developing a male gender. He is currently living as a boy. His interests, manner and play, and refusal to be identified as a girl indicate that M.C.'s gender has developed as male. Indeed, M.C. is living as a boy with the support of his family, friends, school, religious leaders, and pediatrician. . . . Defendants' decision to perform irreversible, invasive, and painful sex assignment surgery was unnecessary to M.C.'s medical well-being. Medical authority recognizes that children like M.C. may be assigned a gender of rearing independent of any surgery, meaning M.C. could have been raised as a girl or a boy until he was old enough for his gender identity to emerge. At that point, M.C. and his guardians could have made appropriate decisions regarding medical treatment—including whether to have any surgery at all.[27]

On the AIC website, Pam Crawford, M.C.'s mother, explains: "By performing this needless surgery, the state and the doctors told M.C. that he was not acceptable or loveable the way he was born. . . . They disfigured him because they could not accept him for who he was—not because he needed surgery. M.C. is a charming, enchanting and resilient kid. We will not stop until we get justice for our son."[28] (John) Mark Crawford, M.C.'s father, added in a different venue: "The real intent of the lawsuit is just to uphold these constitutional principles—integrity of a person's body and some kind of due process for infants where people around them in power are considering doing surgeries like this."[29]

Numerous media sources, from *The Atlantic*[30] to CNN,[31] covered the lawsuit, which also triggered an Internet frenzy across the intersex community. Intersex activists started a Twitter campaign with the hashtag #justice4mc and started circulating an image of M.C. on both Twitter and Facebook, with his back to the camera to protect his privacy (see Figure 6.3).

An article on Counterpunch quoted Curtis Hinkle, the founder of OII (Organisation Intersex International), who spoke to the larger significance of the case: "I personally feel that the tragic mutilation of M.C. reveals the violence perpetrated on all of us, to some degree—by the very act of assigning a gender . . . The legal separation of all humans into just two genders is rooted in a desperate need to impose stereotypical identities on all of us. I have met countless people who have been emotionally mutilated as a result of this socially sanctioned act. . . . Why does the child have to be ordered?"[32] Another news outlet quoted intersex activist (Sean) Saifa Wall, who said, "I speak for the many who cannot speak, living with the shame[,] isolation[,] and secrecy that surround people with intersex conditions. . . . I think it is the time now for us to come together as a community and allow our voice[s] to be heard."[33] Clearly M.C.'s case had struck a chord.

Both the federal and state courts initially denied the defendants' motion to dismiss the case,[34] giving the plaintiffs and the larger community hope that progress will be made toward eliminating medically unnecessary and irreversible surgery designed to erase a child's intersex trait, perhaps to the point of finding it in violation of the U.S. Constitution.[35] However, in late January 2015, the U.S. Court of Appeals for the Fourth Circuit overturned the federal district court's ruling, which means "the federal case will not go to trial."[36] This development is not deterring intersex activists. AIC took a

Figure 6.3. A sign of support. Photo circulated by Advocates for Informed Choice.

strong stance in a statement they released following the decision: "We will not stop fighting for justice for M.C.! We are continuing to pursue the case in state court, and we thank all of our supporters for staying with us on this long road to justice. Progress in fighting for our rights doesn't always happen as quickly as we want. However, no one providing treatment for intersex children *today* could possibly claim that they weren't aware of the serious human and civil rights issues at stake."[37] In many ways, this case is testing law professor Julie Greenberg's theory (2012) that the legal arena may be the best route to protecting intersex children. Greenberg states: "Thus far, the intersex movement has focused most of its efforts on extra-legal strategies. Legal challenges can provide an additional effective tool to improve the treatment of infants and adults with an intersex condition.

Because the current medical protocol for the treatment of infants with an intersex condition may infringe on a child's right to bodily autonomy, a fundamental interest subject to legal protection, legal institutions may be able to play a constructive role in ensuring that the infant's rights are adequately protected" (2012, 128).

The state court has yet to issue a final ruling on the M.C. case, a process that could take years. However, there is at least some evidence that providers are taking the lawsuit, and the intersex advocates behind it, very seriously. Dr. Peter Lee, a co-organizer of the 2005 meeting in Chicago that resulted in the "Consensus Statement on Management of Intersex Disorders" (Houk et al. 2006; Hughes et al. 2006; Lee et al. 2006), has a co-written paper at the *International Journal of Pediatric Endocrinology* that specifically addresses the M.C. case (Lee et al. 2014). A review of intersex medical care since the consensus meeting, the paper notably acknowledges that intersex social movement organizations, like the AIS-DSD Support Group (which is named in the paper) and AIC, are growing in size and power. The paper also acknowledges that although "the term DSD has been widely adopted," it has been contested by some intersex people because of the pathologization *disorder* implies (Lee et al. 2014, 1).

As a result of these recent events, the global intersex community is more visible today than ever before. Whether or not this unprecedented visibility will change intersex medical care is yet to be determined, but this spotlight on intersex is certainly a big step in the right direction, given that powerful medical experts are clearly paying attention. What is also interesting is that under this spotlight, *disorder of sex development* is rarely used. Instead, whether in news reports or court documents, intersex language is almost exclusively employed, perhaps because intersex people who bravely go public tend to be those who view *disorder of sex development* as pathologizing nomenclature, or perhaps simply because it is easier to say *intersex* to the general public. We are thus reminded once again how important it is to be flexible with terminology, so that we can achieve different goals in different realms.

Actions for Liberation

Despite this recent attention on intersex, too much shame, secrecy, and stigma remain associated with intersex traits. My hope is that, regardless of our genitalia and our experiences with intersex, we can continue to fight for liberation from the shackles society rigidly ties to intersex bodies. We can begin to do this, I argue, by directly confronting the gender structure. When we make the gender structure visible—that is, when we understand how gender constrains us at the institutional, individual, and interactional levels of society—we diminish at least some of the powerful control it has over our lives. The gender structure is not a self-constructed, self-regulated black box system. It didn't emerge, nor is it enforced, on its own. Rather, individuals, consciously or not, fuel its workings in any number of ways, even when they pretend it doesn't exist. Gender does not have to be problematic, especially in the arena of gender presentation. What one wears or does to express one's gender identity ought to be insignificant. It is a problem, however, when gendered expectations are imposed onto bodies because of genitalia. In this unfortunately common situation, intersex, trans, genderqueer people, and the like are often the most vulnerable. These expectations, for instance, are what force intersex people, who do not fit neatly into the gender structure, to undergo the medically unnecessary and irreversible surgeries that, as I've shown in this book, may be intended to help but are often quite harmful.

Given my personal, professional, and ethical commitment to the intersex community, I want to conclude *Contesting Intersex* by using the information and insights I have derived from my research to offer seven actions for liberation that intersex activists and our allies can take to decrease intersex stigma and the shame and secrecy that surround it:

(1) Continue to fight for the elimination of medically unnecessary surgeries.
(2) Collaborate with medical allies.
(3) Forge connections across groups in the intersex rights movement with a goal of increasing gender, racial, and class diversity across and within intersex organizations.
(4) Overcome the fear of public exposure.
(5) Engage with formal and informal feminist scholarship.
(6) Recognize that social constructions—most notably sex, gender, sexuality, and medical diagnoses—drive inequalities in our community.

(7) Most important, value the voices of intersex children in the evaluation of intersex medical care.

Action #1: Stop the Surgeries

The first action in a liberatory transformation of the lives of intersex people must involve holding medical professionals responsible for violating medical protocols during the medical management of intersexuality. Although both the 2000 committee statement "Evaluation of the Newborn with Developmental Anomalies of the External Genitalia" and the 2006 "Consensus Statement on Management of Intersex Disorders" are critical of the medically unnecessary surgical modification of intersex bodies (Houk et al. 2006; Hughes et al. 2006; Lee et al. 2006; Committee 2000), these surgeries continue under the guise of protecting children from health risks, especially cancer, even as these cancer claims are inconsistent at best and fallacious at worst.[38] There are no official estimates on the number of intersex people who have been subjected to medically unnecessary surgical interventions. However, my interviews with medical professionals indicate that such procedures are still quite common. Scholars who were studying the community years before I did noted the physical and emotional scarring caused by intersex "normalization" surgery (e.g., Holmes 2008; Karkazis 2008; Preves 2003), yet, as I've documented, these struggles continue to persist throughout the intersex community. I am left asking: Has intersex medical care changed at all in the past decade?

During data collection, I observed one teenager tell other young adults that she didn't want to go through with her scheduled bilateral gonadectomy. This procedure involves the surgical removal of undescended testes, and although it can usually be performed laparoscopically, the consequences, specifically sexual struggles, are long lasting. The teenager had heard about these consequences from older intersex people who had undergone similar procedures. When one peer tried to reassure her by citing the possibility that the surgery would eliminate the risk of gonadal cancer, the teenager asked how often that cancer occurs. When she was told that less than 1 percent of people like her develop cancer,[39] she said, "I still don't know why I have to have surgery. Breast cancer has to be more common and we don't [for preventative reasons] remove breasts." While this teenager's hesitance may have been motivated by fear of surgery, she made a compelling argument. I later learned that her efforts to stop the surgical modification of her body were unsuccessful. Her doctors insisted that surgery was medically necessary to prevent the development of cancer, and her parents obliged. For the sake of this young person and her generation, we must continue and even accelerate our efforts to stop these surgeries.

Action #2: Work with—Not for—Doctors

Liberatory transformation will also require that medical professionals engage in collective collaboration with intersex people and our families while simultaneously calling for us to be open to collaborating with doctors to promote change. Providers must be willing to learn from the intersex community and share at least some of their authority over intersex traits and the intersex body. There is no question that medical professionals who treat intersex traits are trying to help their patients, yet the majority of them are working exclusively from a medicalized body of knowledge and strategies acquired during their medical educations. However, for intersex people to achieve a true liberatory transformation, medical professionals need to extend their scope by engaging with, valuing the knowledge of, and even welcoming the tools of other stakeholders, including people in the intersex community and sociocultural scholars.

True collaboration means more than superficially including intersex people on DSD teams and decision-making committees. Although Cheryl Chase was invited to participate in creating the 2006 "Consensus Statement on Management of Intersex Disorders" (Houk et al. 2006; Hughes et al. 2006; Lee et al. 2006), her official involvement was very limited. As she described it, "At that meeting, surgeons basically sat with their arms locked and bullied everybody." This is simply not acceptable. Furthermore, parents of intersex children should also be included in meaningful ways, especially on DSD teams where they can witness the medical uncertainty that often happens behind the scenes. It is important to include sociocultural scholars as well, as they bring to the table different views about sex, gender, and sexuality from those of most medical providers. For example, doctors should not only include sociocultural scholars on DSD teams but also take their ideas seriously in order for them to have meaningful input in the group. This extreme hesitance to engage with sociocultural scholars might be due to any number of reasons, but it needs to stop.

Ultimately, I trust that the passion and commitment medical providers have for helping others will eventually draw them to real collective collaboration. I also have faith that our intersex community can move toward working with doctors—as opposed to against them—in ways that will benefit future generations of intersex people and our families. The medical profession is a powerful institution. It is in our interest to work with that power to promote change. With supportive medical allies as our secret weapons, we can work together to change intersex medical care.

Action #3: Expand and Diversify Peer Support

Liberatory transformation also depends upon expanding and diversifying peer support and the informational biocitizenship such support enables. When I asked research participants what advice they would give to someone recently diagnosed with an intersex trait, almost everyone said they would strongly encourage that person to meet other intersex people. Intersex people, me included, find solace in knowing we are not alone in our experience of being differently bodied. My interviews with parents confirmed the value of peer support. Parents with whom I spoke reported that meeting other parents of intersex children was emotionally empowering.

This might be one of the simplest of the seven actions for liberation to employ, given that today the newly diagnosed can access intersex organizations on the Internet. As Sharon Preves (2003) reported more than a decade ago, many intersex people find similarly bodied individuals through the Internet—a fine example of digital biocitizenship—and the practice can only have increased in recent years. Although not everyone has access to the Internet, between smartphones and libraries that access itself has increased dramatically.[40]

Given the importance of peer support, I strongly recommend that all intersex groups strive toward being more inclusive of all intersex people, regardless of their gender expression. In fact, this is one of my goals as president of the AIS-DSD Support Group, where I want to make sure all intersex people are welcomed. Although AISSG-USA changed its name in 2011 to the AIS-DSD Support Group to be more inclusive to those with intersex traits other than androgen insensitivity syndrome (AIS), the renaming created controversy across the global intersex community because it included the abbreviation *DSD*, which stood for *disorders* of sex development. In the fall of 2014, the AIS-DSD Support Group unveiled a redesigned website on which "*disorders* of sex development" terminology was, in the majority of instances, replaced with "*differences* of sex development" language.[41] I personally hope this move, among many other changes we've made as an organization (see chapter 2), makes us more inviting to all intersex people. There also needs to be more collaboration across intersex organizations toward a shared goal of increasing racial and class diversity throughout the global intersex community. As it stands today, the intersex community consists largely of individuals privileged by race and class. We must actively work to diversify our community.

Action #4: Replace Fear with Power

In order to achieve liberatory transformation, we must replace the fear of not fitting into the sex binary with the power we gain by educating others about sex variability. In my interviews with intersex people, I saw that the individuals who are most emotionally comfortable with their intersex diagnosis are those who do not feel constrained by their difference but rather embrace it as part of their identity. Many of these individuals speak publicly about their lived experiences in venues as varied as daytime television talk shows,[42] *The New Yorker*,[43] hospitals and medical conferences, and universities across the country. Their presentations range from descriptions of the unwanted surgical interventions they were subjected to as children to answering questions about what it is like to be differently bodied. This public exposure seems to be related to both a positive sense of self for these individuals and a destigmatization of intersexuality that benefits the entire intersex community. Given how important the media have been for the intersex community, I agree with Stevie's proclamation that there needs to be "more public exposure . . . just getting the word out . . . getting more airtime on television, and other forms of media."

Action #5: Embrace Feminist Ideas

If we believe, as my research suggests, that ideas can themselves be liberatory, and classrooms can be spaces where these liberatory ideas are shared (e.g., hooks 1994), then another step we must take on our road to liberatory transformation is to embrace the insights of feminist scholarship. Exposure to feminist scholarship created strikingly different outcomes for my research participants. Individuals who encountered feminist scholarship in formal educational settings or were introduced to it by friends or the media reported much more positive senses of self and were much more comfortable with their intersex trait than those who weren't. Twenty-two-year-old Eve was enrolled in a humanities seminar whose students were required to write a seminar paper on a topic of their choice. One of her classmates chose to write about intersex. Having never shared with any classmates or instructors that she had an intersex trait, Eve initially felt uneasy about having a fellow student study intersex:

> It seemed like it was going to be really interesting, because she was going to write about the feminist movement and intersex, and everyone was really interested in it, but I was sitting there in the classroom kind of thinking to myself, God, am I really just going to sit here for like four months and let people talk about me and not say something? So one day, I was just like okay, I worked up my courage and I ran after her after class and I was like, "Hey, I just have to tell you, *I'm* intersex and I have AIS." She's like, "Oh, okay, that's cool. Yeah, I think I read about AIS." So we ended up talking for like 15 minutes. I just . . . I knew that I needed her to know so that she would like not be more sensitive, but just maybe be like aware of the fact that there was someone in the room that was really affected by everything that people were talking about. I don't know. It was just important to me to not be silent . . . it felt really good to share it with her after it was done.

Direct exposure to feminist scholarship in a classroom setting isn't the only path to feminist scholarship. In fact, far more individuals I spoke with cited the media as the source of their introduction to a feminist lens on intersexuality. Liz, a thirty-three-year-old woman, said she first sought out others like herself after seeing an episode of *Oprah* titled "Growing up Intersex,"[44] which featured a panel of three individuals with different intersex traits and historian Alice Dreger, who has written extensively on the topic of intersexuality (see, e.g., Dreger and Herndon 2009; Dreger 1998a,b,c).

When I interviewed forty-three-year-old Vanessa in October 2009, she was uncomfortable with intersex language and expressed preference for DSD terminology. She was new to the intersex community. When I asked her what lay at the core of her frustration with the intersex diagnosis, she said, "It's not the [lack of] reproduction [that we can't have biological children]. . . . It's more of the actual identity. . . . It's more of being different. It's the issue of being different and not fitting into a category." Several months later, I found myself on an airport bus with Vanessa on our way to a four-day social gathering of intersex people. Vanessa inquired about the progress of my project and in the course of our conversation pulled Fausto-Sterling's (2000a) *Sexing the Body* from her bag. She told me that one of our mutual friends had suggested she read it, and while she just wanted to know whether I was familiar with the book, it was immediately clear that she had a different sense of herself from when I had interviewed her several months earlier. She appeared more comfortable with her intersex trait, which leads me to believe that connecting with others had transformed her, especially those exposed to feminist ideas. She now regularly uses intersex language and has even moved into a leadership role in the community.

At an AISSG-USA support group meeting in the summer of 2010, I found myself with thirty other women in a private session designed for intersex people to share their experiences, frustrations, and concerns. Having attended this session at past meetings, I knew it was the most emotionally charged few hours of the entire conference, and its mood was always somber, as all participants had the opportunity to share anything on their minds. This time, when it was my turn to share with the group, I felt a sudden urge to change the mood of the room by sharing an experience that called for what I see as a feminist sense of humor. I told them that as an undergraduate, I participated in market research studies for extra money. They paid quite well and covered an array of topics ranging from pet food to pizza to toothpaste. During one particular study, I sat around a conference room table with six other women, including the facilitator. The walls of the room were covered with one-way mirrors, and multiple microphones dangled strategically from the ceiling to capture our every word. When we first entered the room, I chose to sit next to the focus group facilitator. Throughout all of my classroom experiences, I've always chosen to sit in the seat closest to the instructor, so this wasn't new. When everyone was comfortably situated, the facilitator informed us that we would be discussing sanitary napkins. Having been born with androgen insensitivity syndrome, I had never had a menstrual period, but I figured I would do a good enough job pretending in order to get the $150.00 compensation I had been promised for my feedback. The facilitator began by asking me how many pads I use in a typical cycle. I must have looked like a deer in headlights when I responded, "Four." The look on everyone's face can only be described as a strange combination of shock and confusion. I quickly modified my answer, "a day?" With their faces unchanged, I again modified my answer: "On a light day?" I asked.

As I shared this story in the private session, the room dramatically transformed from a somber space to one filled with laughter, to the point that some were even brought to tears. I told everyone that I found solace living as I am. I recognized my own agency in replacing stigma with knowledge of the social construction of not only sex, gender, and sexuality but also medical diagnoses (see Action #6). It was then I realized that the theoretical frameworks I had been exposed to throughout my sociological training made this easier for me than it often had been for others. We all need the benefits of these insights, so we can better understand the nature of our struggles and challenges.

Action #6: Recognize Social Constructions

This sixth action for a liberatory transformation builds on the previous action and calls for us to understand that phenomena throughout the world are constructed by various institutions in ways that maintain and perpetuate inequalities. The medical profession is just one example of such an institution. In the intersex community, the most debilitating phenomena include sex, gender, sexuality, and medical diagnoses, all of which tend to be viewed as indisputable characteristics of the body or, in the case of diagnoses, indisputable descriptions of those characteristics.

It is not common to recognize sex, gender, sexuality, and even medical diagnoses as socially constructed through various social institutions, but doing so could be liberatory for the intersex community.

How we understand, for example, our sex anatomy and the rather arbitrary differences between male and female bodies is taught to us by medical providers, sex education curricula, and the media, among other institutions. As sociologist Asia Friedman (2013) has argued, this socialization leaves us blind to the sameness between "male" and "female" bodies. At the same time, it equips us to search for the characteristics that enable sex categorization. Interestingly, this search is not limited to our visual senses, as Friedman found this same process was at play in her subsample of blind participants. Our understanding of sex as a biological phenomenon is not natural. Rather, it is reinforced by various institutions that are often left unchallenged and positioned, consciously or not, as authorities on the subject matter. This is not to reify institutions by suggesting they have a life of their own outside of human action. The medical profession, as an institution, comprises medical providers who are in most cases subject to the same narrow ideologies about sex, gender, and sexuality all of us encounter throughout our lives. Recognizing social constructions would allow the intersex community to see how medical providers and others draw on other ideologies, including narrow ideologies about sex, gender, and sexuality, when they define (and even treat) diagnoses. This shouldn't be a difficult assumption for the intersex community to accept and ultimately employ in search of liberation. After all, we just need to consider how definitions of intersex have, themselves, been routinely contested throughout our community's history to understand how this powerful dynamic works.

Action #7: Listen to Children

I see the seventh and final action as one of the most important. We must respect and include children's voices in the medical decision-making process. I recognized the importance of this action after reading sociologists Monica Casper and Lisa Jean Moore's book *Missing Bodies: The Politics of Visibility* (2009). In their book, Casper and Moore suggest that children are missing from sexuality research because of an Institutional Review Board (IRB) process that is "both insufficient and overkill"

(54), as well as a cultural attitude in which children "are typically preserved in an ideal type of 'innocence' that may not be, and quite often is not, reflective of their reality" (28). While Casper and Moore focus on children's sexuality, their insights are also applicable to children in the intersex community.

More often than not, medical providers ignore intersex children. That is, children's voices aren't usually heard in both specific decisions about their own bodies or in larger debates about intersex medical care, despite the fact that they are the ones most affected by the medical management of intersex traits. In most cases, medically unnecessary surgical interventions on intersex bodies happen to children—that is, to minors who cannot legally make their own decisions, or, if they are babies or toddlers, cannot even formulate their own opinions. Many medical providers who treat intersex children do not seriously consider children's wishes in the medical decision-making process. In order to value children's voices, medical providers would need to avoid all medically unnecessary interventions until children were mature enough to make their own decisions about their bodies. Given that these irreversible procedures are more often than not cosmetic, shouldn't intersex children have the right to delay or refuse them? I'm not suggesting that all medical recommendations pertaining to a child's intersex trait should be withheld until the child is mature enough to assess them. Rather, I'm suggesting that in the case of medically unnecessary and irreversible surgical interventions, we need to listen to intersex children and involve them in the decision-making process.

Medical providers aren't the only ones not listening to children. Sociocultural scholars have also narrowly focused their studies of the intersex community on intersex adults. I wonder why no sociocultural scholars have thought to ask intersex children in a systematic fashion how they feel about their intersex diagnoses. While it might be difficult to receive IRB approval to study children, given their protected research status, this challenge should not preclude us from trying. Some people may believe that intersex youth would feel something was wrong with them if they were invited to participate in research studies, but isn't it possible to interview kids in a way that's empowering rather than stigmatizing? Sociologist Margaret Hagerman (2010) conducted child-centered interviews with twenty kids between the ages of five and twelve and found her participants to be engaged in the research process and generally to enjoy it. Although Hagerman was not interviewing intersex kids, her findings invite us to question the assumption that we are protecting children by excluding them from our studies. Our failure to include youth voices in our work denies intersex kids a space in which to share their intersex experiences confidentially. Although we do not include children in a meaningful way in research studies on intersexuality, they are the ones most directly affected by medical protocols in the short and long term. As it stands, we are failing to understand the complexities of the intersex community by excluding their voices.

Although the voices of intersex children are not included in medical protocols, decision-making processes, or sociocultural studies, children are definitely active in the intersex community. In 2010, with a generous grant from the Ms. Foundation for Women (and subsequent funding from the Liberty Hill Foundation), AIC created Inter/Act, the youth advocacy program introduced earlier in this chapter. AIC describes Inter/Act as "the first and only intersex youth leadership and movement-building group in the country,"[45] and it is indeed an incredible program that offers youth with intersex traits the spotlight they have never had. In 2012, AIC started a letter-writing campaign for youth to share what they wished their doctors knew. They then produced and distributed a brochure with excerpts from the letters. One young person said, "I know plenty of people who've had poor experiences with intersex surgeries. Doctors need to be honest about the frequency of failures and complications."[46] Another said, "After many years of complications, I was referred to a doctor who was one of the finest people I've ever met. She cared for me. She would sit with me for an hour and just talk. . . . What was I feeling? What did I want? What worked and what didn't? It made all the difference in the world."[47]

In 2013, AIC appointed Pidgeon Pagonis as the new Youth Leadership Coordinator of Inter/Act. That year, Inter/Act also joined forces with the AIS-DSD Support Group to collaborate on programing for youth who attended the annual AIS-DSD Support Group meeting. I witnessed first-hand how this collaboration, which highlighted youth voices, was liberatory for the children and young adults in our community. They walked away with screen-printed t-shirts they had made (see Figure 6.4), new friends, and a positive outlook on what it means to be born with an intersex trait in today's society. In 2014, Inter/Act hosted the first ever youth intersex retreat, collaborated again with the AIS-DSD Support Group at their nineteenth annual meeting, and even presented to a room full of medical providers at the pre-conference Continuing Medical Education (CME) meeting. Intersex youth are attempting to shape dominant discourses concerning intersex medical care. We must listen to them.

Conclusion

When I began my research in the intersex community, I had no idea that I would eventually tell the story of how *intersex* became a *disorder of sex development* and the subsequent conflicts over the latter term. I also never would have predicted that I would move away from my insider/outsider role to an obvious insider position in the U.S. intersex community. When I was elected president of the AIS-DSD Support Group, it was clear that I had assumed a position of significant power in the intersex community.

I suspect my newly acquired insider status will raise methodological concerns for some readers of *Contesting Intersex*. People may wonder how I balanced my desire to produce a complete and accurate analysis of the intersex community with my own participation in that community. How did I remain true to my personal and professional standards of ethical research practices? How did I present an unbiased view of the intersex community, given my transition from insider/outsider status to leader of one of the most well-known intersex organizations in the world? My answer to these questions, and others like them, is that research is always biased on the basis of any number of factors, including our personal biographies. My personal experience with intersex appears throughout *Contesting Intersex*, which may make the book read like an autoethnography at certain points. And, I'm okay with that. As sociologist Sara Crawley eloquently put it, "Scholarship, even autoethnography, is best understood as part of the group conversation we call knowledge construction" (2012, 156).

Since my first contact with the intersex community, I have grown in multiple ways and built numerous friendships with intersex people, their parents, medical providers, and community allies. I no longer feel

Figure 6.4. The kids are (dressed) all right. Photo by Pidgeon Pagonis.

ashamed, stigmatized, or deceived by medical providers about my intersex trait. I feel liberated and fortunate to have been born with an intersex body that automatically afforded me membership in our beautiful community. However, our community's work is far from over. Medical providers are still performing medically unnecessary and irreversible surgeries on intersex children, while many intersex adults continue to struggle with feelings of abnormality. Intersex people should not have to navigate their diagnoses alone, but many do because providers are reluctant to refer their patients to support groups despite substantial evidence from sociocultural scholars about the emotional benefits of connecting with similarly situated others.[48]

My hope is that the research I offer in this book provides a refreshingly unique take on the sociocultural study of intersex and, more broadly, sex, gender, and sexuality binaries. On a more personal level, I hope the actions I've outlined in this chapter will be beneficial to our community and useful to other groups facing similar struggles. I've titled this book *Contesting Intersex: The Dubious Diagnosis* not because intersex isn't real—I know first-hand that it is—but because it is only as real as its definition and the mechanisms we use to enforce such medical labels. One doesn't need to be a linguist to see how the naming of our traits has an unsteady history. In 1997, my medical records indicated that I was a male pseudohermaphrodite with testicular feminization syndrome. A few years later, medical providers recorded in my chart that I had an intersex condition and complete androgen insensitivity syndrome. Today, they note that I have a disorder of sex development. If I had been born in 1780 instead of 1980, my intersex trait might have gone undiscovered. After all, my body is no different from any other woman's body, although I have less pubic, underarm, and leg hair. It is probable that I would have been diagnosed with amenorrhea (the absence of menstruation), but unless invasive internal physical examinations had been conducted, it is very unlikely that my intersex trait would have been discovered. These examples make it clear that, regardless of what terminology one evokes, intersex is a dubious diagnosis.

I don't know if intersex language will be around much longer, but neither do I know how long DSD terminology will be around. What I do know is that a diverse community of people with intersex traits will continue to exist, and, as long as I am able, I will be there to welcome these individuals into our global intersex community in such a way as to simultaneously disrupt the assumptions that diagnoses are real and that sex, gender, and sexuality are natural, binary, and correlated characteristics of the body. We're intersexy and I know it . . . and I hope now you do, too.

APPENDIX A: TABLE OF RESEARCH PARTICIPANTS

In Table A.1, I display demographic characteristics for each of my participants. In order to protect confidentiality, I only offer a limited amount
of information about each participant.

TABLE A.1. Select Demographic Characteristics of Participants (N = 65)

Participant	Background[1,2]	Gender Presentation[3]	Age	Self-Reported Sexuality	Education Status[4,5]
Aimee	Individual	Woman	30	Heterosexual	College educated
Alexis	Parent	Woman	48	Heterosexual	Some college
Alice	OBM	Woman	Unknown	Straight	College educated
Ana	Individual	Woman	37	Homosexual	Some college
Ann	Individual	Woman	52	Gay	College educated
Bruce	Individual	Man	40	Complicated	College educated
Caitlin	Individual	Woman	26	Queer	Didn't finish HS
Cheryl	Individual	Woman	53	Lesbian	College educated
Chris	Individual	Man	57	Asexual	College educated
David	Individual	Man	62	Asexual	College educated
Donna	Individual	Woman	53	Lesbian	College educated
Drew	Parent	Man	39	Heterosexual	College educated
Dr. A	Doctor	Man	61	Heterosexual	College educated
Dr. B	Doctor	Woman	Unknown	Heterosexual	College educated
Dr. C	Doctor	Man	Unknown	Heterosexual	College educated
Dr. D	Doctor	Woman	53	Heterosexual	College educated
Dr. E	Doctor	Woman	54	Lesbian	College educated
Dr. F	Doctor	Woman	54	Heterosexual	College education
Dr. G	Doctor	Woman	Unknown	Heterosexual	College educated
Dr. H	Doctor	Man	Unknown	Complicated	College educated
Dr. I	Doctor	Woman	Unknown	Heterosexual	College educated
Dr. J	Doctor	Man	59	Homosexual	College educated
Emily	Individual	Woman	50	Bisexual	College educated
Esther	Individual	Woman	53	Lesbian	College educated
Eve	Individual	Woman	22	Complicated	College educated
George	Parent	Man	51	Heterosexual	College educated
Hannah	Individual	Woman	42	Lesbian	College educated
Hope	Parent	Woman	40	Heterosexual	Some college
Irene	Individual	Woman	72	Bisexual	College educated
Jane	Individual	Woman	54	Heterosexual	Some college
Jeanne	Individual	Woman	49	Bisexual	College educated
Jeff	Parent	Man	50	Heterosexual	College educated
Jen	Parent	Woman	36	Heterosexual	Unknown
Jenna	Individual	Woman	31	Straight	College educated
John	Parent	Man	35	Heterosexual	College educated
Karen	Individual	Woman	52	Straight	College educated

Kelly	Individual	Woman	38	Heterosexual	College educated
Kimberly	Individual	Woman	38	Lesbian	College educated
Laura	Parent	Woman	29	Heterosexual	College educated
Leigh	Individual	Woman	24	Queer	College educated
Liz	Individual	Woman	33	Heterosexual	College educated
Maria	Individual	Woman	32	Bisexual	College educated
Mariela	Individual	Woman	29	Straight	Some college
Marilyn	Individual	Woman	50	Straight	College educated
Marty	Parent	Woman	43	Heterosexual	Some college
Mercurio	Individual	Gender-Fluid (self-identified)	41	Lesbian	College educated
Michael	Parent	Man	61	Heterosexual	College educated
Michelle	Parent	Woman	42	Heterosexual	Some college
Millarca	Individual	Woman	46	Lesbian/Dyke	College educated
Paul	Individual	Man	57	Asexual	College educated
Peggy	Individual	Woman	56	Heterosexual	College educated
Pidgeon	Individual	Genderqueer (self-identified)	23	Queer	College educated
Rachel	OBM	Woman	40	Straight	College educated
Rebecca	Individual	Woman	35	Lesbian	College educated
Rosland	Individual	Woman	65	Heterosexual	College educated
Sarah	Parent	Woman	55	Heterosexual	HS graduate
Shelby	Parent	Woman	38	Heterosexual	College educated
Sheryl	Parent	Woman	49	Heterosexual	Some college
Skywalker	Individual	Woman	37	Heterosexual/ Bisexual	College educated
Steve	Parent	Man	55	Heterosexual	Less than HS
Stevie	Individual	Woman	44	Queer	College educated
Sue	Parent	Woman	51	Heterosexual	College educated
Susan	Parent	Woman	57	Heterosexual	College educated
Tara	Individual	Woman	23	Heterosexual	College educated
Vanessa	Individual	Woman	43	Heterosexual	College educated

1 "OBM" refers to an organizational board member who doesn't have an intersex trait/DSD or isn't the parent of someone who does.

2 Two of the medical professionals I spoke with didn't have either an MD or PhD degree, yet I refer to them as "Dr." throughout in order to ensure confidentiality.

3 Unless otherwise noted, gender presentation is based on my perception given hegemonic U.S. cultural cues.

4 "Some college" refers to the participant's having completed some college that did not result in a degree.

5 "College educated" refers to the participant's having completed at least a four-year degree.

Chicago: October 27–31, 2005

Organisers: Ieuan Hughes (ESPE) and Peter Lee (LWPES)

FACULTY AND SUMMARY OF QUESTIONS TO BE ADDRESSED

GROUP 1 Recent Molecular Genetic Impact of Human Sexual Development
Co-ordinators: Olaf Hiort (Germany) and Eric Vilain (USA)
Members: Jean Wilson (UT Southwest, Dallas, TX)
              Vincent Harley (Prince Henry's Institute of Med
                  Research, Melbourne, Australia)
              John Achermann (London)
              Erica Eugster (Indiana University)

QUESTIONS:

1. What is the current state of knowledge of molecular mechanisms of sexual development in humans? Are there "activational" and/or "organisational" effects?
2. What lessons can be learned from animal models?
3. Should there be a new nomenclature for the classification of intersexuality based on genetic etiology?
4. What is the current availability of genetic testing for intersexuality, including prenatal testing. Should there be "Centres of Excellence" for testing?
5. Is there any genotype/phenotype correlation that would provide clinically useful information?
6. Are there ethnic or geographic influences on the prevalence of specific intersex conditions? (see also Question 3 in Group 6).

GROUP 2 Brain Programming by Genes and Hormones (evidence-based)
Co-ordinators: Melissa Hines (UK) and Sheri Berenbaum (USA)
Members: Peggy Cohen-Kettenis (Netherlands)
            Jay Giedd (USA)
            Anna Nordenström (Sweden)
            Bill Reiner (USA)
            Emilie Rissman (USA)

QUESTIONS:

1. What are the causes of gender identity disorder in individuals without intersex conditions?
2. What are the causes of gender dysphoria and gender change in individuals with intersex conditions?
3. What are the human behavioural effects (excluding gender identity) of prenatal androgens and what characteristics of hormone exposure account for variations across individuals and across behaviours? How are behavioural effects of androgens separable from physical (especially genital) effects?
4. How are neural and behavioural effects of early hormones in rodents and primates dependent on characteristics of hormone exposure and social context? How are behavioural effects of androgens separable from physical (especially genital) effects?
5. What is the role of the sex chromosomes in behavioural masculinisation and feminisation in human and non-human species?
6. How does brain structure differ in human males and females at different stages of development?

GROUP 3 Investigation and Medical Management of Intersex in the Infant, Child, and Adolescent
Co-ordinators: Pat Donohoue (USA) and Faisal Ahmed (Scotland)
Members: Sylvano Bertelloni (Italy)
            Felix Conte (USA)
            Claude Migeon (USA)
            Chris Driver (Scotland)
            Kenji Fujieda (Japan)
            Cheryl Chase (USA)

QUESTIONS:
1. What are the definitions of normal and ambiguous genitalia?
2. Description of the evaluation of patients with ambiguous genitalia, including those detected later than the newborn period.
3. What are the factors influencing the choice of sex of rearing?
4. Medical management in cases of gender re-assignment.
5. General description of counselling to be offered to parents of newborns with ambiguous genitalia (see also Question 1, Group 5).
6. Management of the adolescent with an intersex condition, with attention to fertility concerns.

GROUP 4 Surgical Management of Intersex

Co-ordinators: Laurence Baskin (USA) and Pierre Mouriquand (France)
Members: John Brock (USA)
              Rick Rink (Riley Children's Hospital, USA)
              Mel Grumbach (UCSF Children's Hospital, USA)
              Phillip Ransley (England)

QUESTIONS:
1. Clitoris: should surgery be performed on the clitoris?
2. Vagina: surgical management of the common urogenital sinus. Timing and technique.
3. Vagina: vaginal substitution: who, how and when?
4. Penis: reconstruction, durability, hormonal stimulation and tissue engineering.
5. Gonads, Wolffian and Mullerian Structures: cancer risk, removal and timing (see also Question 4, Group 6).
6. Surgical factors influencing the gender assignment in the "Y deficient" patient.

GROUP 5 Psychosocial Management of Patients with Intersexuality and Related Conditions

Co-ordinators: Heino F.L. Meyer-Bahlburg (USA) and Polly Carmichael (England)

Members: David Sandberg (USA)
                   Froukje Slijper (Netherlands)
                   Norman Spack
                   Barbara Thomas (Germany)
                   Kenneth Zucker (Canada)

QUESTIONS:

1. Gender assignment: how should gender assignment be psychosocially managed in newborns?
2. Gender re-assignment: how should gender re-assignment be psychosocially managed in children, adolescents and adults?
3. Genital surgery and sex-hormone treatment: how should the psychosocial aspects of genital surgery and sex-hormone treatment be managed?
4. Information management: how should disclosure of sensitive personal information be handled?
5. Sexuality: how should intersex-related problems in romantic and sexual functioning and orientation be handled in adolescents and adults?
6. Structural issues: how can the need for clinical collaboration of multiple disciplines be accommodated?

GROUP 6 Outcome Data: Evidence-based
Co-ordinators: Sten Drop (Netherlands) and Garry Warne (Australia)
Members: B Mendonca (Bra)
                   L Looijenga (Netherlands)
                   U Thyen (Denmark)
                   J Schober (USA)
                   A Wisniewski (USA)

QUESTIONS:

1. What constitutes "long-term" outcome? Which parameters should we use to measure physical and psychosexual long-term outcome, both subjectively and objectively, for the purposes of this study?
2. What is the long-term outcome per diagnostic category with emphasis on adulthood and type of surgical treatment?

3. What is the role of culture and social circumstances on the long-term outcome?

4. What is the incidence and what are the diagnostic criteria regarding gonadal tumours?

5. What is the impact of repeated medical examination, photography and surgery on the occurrence of the development of the psychological conditions?

6. Are there additional health problems in any of the diagnostic categories of intersex?

NOTES

CHAPTER 1. INTRODUCTION

1 For a theoretically rich and extensive discussion of the origins and controversies over sex chromosomes, see Sarah Richardson's (2013) *Sex Itself*.

2 To quote Sharon Preves (2001), "Interestingly, the term *intersex* emerged in the late nineteenth century and was used not only when referring to hermaphrodites but to homosexuals as well (Epstein 1990)" (523).

3 DSD terminology was first introduced in "Changing the Nomenclature/Taxonomy for Intersex: A Scientific and Clinical Rationale" in the *Journal of Pediatric Endocrinology & Metabolism* (Dreger et al. 2005). The DSD abbreviation stood for something slightly different in this paper: disorders of sexual differentiation.

4 Several others (e.g., Topp 2013 and Spurgas 2009) have written about nomenclature divides in the intersex community, but my approach is different in that it provides a thorough empirical analysis of these divides, built on in-depth interviews with intersex people, their parents, medical experts, and other intersex advocates.

5 Cited from Sharon Preves's (2003) *Intersex and Identity*, p. 27.

6 This estimate of known intersex traits in the population comes from a handbook that was originally published in 2006 by ISNA but is now distributed by Accord Alliance after copyright was assigned to the new organization in 2008. The handbook is titled *Clinical Guidelines for the Management of Disorders of Sex Development in Childhood*. It was written by the Consortium on the Management of Disorders of Sex Development (see Consortium 2006a).

7 Katrina Karkazis (2008) raises the issue in her book *Fixing Sex*.

8 "Peggy Cadet," a person in the intersex community, has been extremely critical of widely circulated intersex statistics. See also Peggy (2014) and Peggy (2013).

9 For a further discussion of estimates of intersex in the population, see Melanie Blackless et al. (2000) and Anne Fausto-Sterling (2000a, 53).

10 We now know that this claim is not empirically supported. See, for example, Rola Nakhal et al. (2013), Rebecca Deans et al. (2012), J. Pleskacova et al. (2010), Leendert H.J. Looijenga et al. (2007), Martine Cools et al. (2006), and Deborah Merke and Stefan Bornstein (2005).

11 More details about the John Money fiasco are presented in chapter 3. See also John Colapinto (2000, 1997).

12 One of the medical professionals I interviewed identified as having an intersex trait.

13 This is also an example of digital biocitizenship (see Rose and Novas 2005; Rose 2007, 2001).

14 Brian Still (2008) offers an in-depth discussion of the Internet's role in the intersex community in his book *Online Intersex Communities: Virtual Neighborhoods of Support and Activism*.

15 Candace West and Don Zimmerman first described this in 1987 in their *Gender & Society* publication "Doing Gender."

16 See Costello (2015a).

17 See Costello (2015b).

18 For more discussion of standpoint theory, epistemology, and methodology, see Joey Sprague (2005), Nancy Naples (2003), Sandra Harding (1998), and Nancy Hartsock (1983).

19 See, for example, Robert Groves et al. (2004).

20 See, for example, Carolyn Chew-Graham, Carl May, and Mark Perry (2002).

21 See, for example, Robert Groves et al. (2004).

22 For a historical discussion of the divides in the intersex community, see Sharon Preves (2005).

23 In "Medicalization and Social Control," Peter Conrad (1992) defines medicalization as a "process by which nonmedical problems become defined and treated as medical problems, usually in terms of illnesses or disorders" (209).

24 See Candace West and Don Zimmerman (1987) for a discussion of "doing gender."

25 See "Project Integrity."

CHAPTER 2. THE TRANSFORMATION OF INTERSEX ADVOCACY

1 Phil Brown et al. (2004) describes embodied health movements in great detail.

2 Social movement literature has long wrestled with the necessity of a collective identity for successful social movement organizing (e.g., Dugan 2008; Gamson 1995; Ghaziani 2008; Hunt and Benford 1994; Polletta and Jasper 2001; Reger, Myers, and Einwohner 2008; Rupp and Taylor 1999; Taylor and Whittier 1992; Taylor 2000; Valocchi 1999).

3 It is not unheard of for a movement to adopt, move away from, or return to a collective identity. As sociologist Mary Bernstein (1997) has argued, activists either promote an identity of similarity with those outside of the movement to gain public support, or they construct an identity of difference whereby they exemplify their difference from the general public. The collective identity constructed by activists can shift, especially in response to political changes.

4 Sociologist Barbara Risman introduced the idea of gender as a social structure in *Gender Vertigo* (1998). She further developed gender structure theory in 2004 in an article in *Gender & Society*. The idea that gender is central to social movements is not new. Sociologist Verta Taylor (1999, 1996, 1995, 1989) has written extensively on the relationship between gender and social movements. See also Leila J. Rupp and Verta Taylor's (2003) *Drag Queens at the 801 Cabaret* and Verta Taylor and Nancy Whittier (1992).

5 J. David Hester notes that intersex bodies "raise a threat to gender" (2004, 223).

6 See Nikolas Rose and Carlos Novas (2005) for a discussion of the biosocial community.

7 It is clear that mobilization strategies are fundamental to social movement organizations, though different organizations may employ very different tactics, some of which hardly look like tactics at all. Mary Fainsod Katzenstein (1990), for example, argues that women activists within the Catholic Church and U.S. military in the 1980s and 1990s engaged in "unobtrusive mobilization" to promote change and raise "gender consciousness" by working within institutions as opposed to engaging in a more public and confrontational mobilization strategy. Patricia Yancey Martin's (2005) study of rape crisis centers offers further evidence to support the success of "unobtrusive mobilization." Rape crisis centers are successful in gaining support, resources, and attention by not "stand[ing] outside and [allocating] blame" but rather by employing "unobtrusive mobilization" by "persuad[ing] outsiders to adopt their versions of laws, police officer training, rape exams, and school health education messages" (Schmitt and Martin 1999, 379; also see Martin 2005).

8 See Sarah Richardson (2013).

9 Debates having to do with naming, defining, and treating medical diagnoses are not unusual (see Brown 1995; Conrad 2007; Cooksey and Brown 1998). Consider the case of attention deficit hyperactivity disorder (ADHD). According to sociologist Peter Conrad (2007), the expansion of the ADHD diagnosis to include adults in the 1990s was met with criticism. The Church of Scientology, for one, publicly criticized the diagnosis, and some therapists also expressed a concern that the ADHD diagnosis was "becoming too prevalent" (Conrad 2007, 60). Disputes surrounding the *Diagnostic and Statistical Manual of Mental Disorders* (*DSM*) provide another example. While psychiatrists embrace the *DSM*, "criticism . . . comes from social workers, psychologists, and others for whom it does not foster professional dominance" (Cooksey and Brown 1998, 549). Feminist critiques of the treatment of intersex traits therefore stand in good company.

10 As noted by Sharon Preves, "Dreger's [1998c] historical research shows that British medical doctors George F. Blacker and T. W. P. Lawrence originated this five-sex system in 1896" (2001, 539).

11 Anne Fausto-Sterling states this in her article "The Five Sexes, Revisited" (2000b, 19).

12 Myra Hird (2000) also discusses the social construction of sex in her article titled "Gender's Nature: Intersexuality, Transsexualism and the 'Sex'/'Gender' Binary."

13 The National Organization for Women also passed a resolution in 2001 that condemned intersex medicalization. See "NOW: Leading the Fight."

14 This quote comes from Chase's piece titled "Making Media: An Intersex Perspective," which first appeared in GLAAD *Images*, Fall 1997. It was later republished by GLAAD on May 1, 2002, and shared with me by Chase, as it is no longer publicly accessible.

15 We will likely continue to see the importance of the Internet in the intersex community as we grow increasingly connected through social networking sites such as Facebook.

16 See "FAQ" (2008).

17 The decentralization of OII occurred alongside technical changes. When I was in the field, OII was housed on the Internet at www.intersexualite.org. I recruited from this OII website and involved any of its members or affiliates interested in participating in my study so long as they identified as U.S. citizens. Today, the home base for OII, or OII Intersex Network as it is currently named, is housed at http://oiiinternational.com/. From there, one can connect to any of OII's twelve different global affiliates.

18 See Admin (2012).

19 See Hinkle and Viloria (2012).

20 This excerpt comes from Cheryl Chase's 1997 "Making Media: An Intersex Perspective" GLAAD *Images* publication.

21 See Hinkle and Viloria (2012).

22 We can think of the intersex rights movement, especially 1990s intersex advocacy, as an embodied health movement with a goal of contesting the framing of intersex as an illness.

23 The first newsletter was titled "Hermaphrodites with Attitude." It was distributed in the winter of 1994.

24 See Janik Bastien-Charlebois and Vincent Guillot (2013) for a critical discussion of the ways in which some medical experts confrontationally approach intersex activists. This research contribution was published in French. Because I do not read French, Bastien-Charlebois graciously offered a translation.

25 Androgen insensitivity syndrome was once labeled testicular feminization syndrome.

26 In 2010, the Lawson Wilkins Pediatric Endocrine Society was formally renamed the Pediatric Endocrine Society. See "History of the Society."

27 It should be noted that infertile heterosexuals are still understood as appropriately gendered.

28 This excerpt comes from Cheryl Chase's 1997 "Making Media: An Intersex Perspective," GLAAD *Images* publication.

29 Alice Dreger also discusses on her website the reasons she advocated for DSD terminology. See Dreger (2007).

30 The agenda is included in Appendix B and was obtained and included with permission from the U.K. AISSG group.

31 See "DSD Terminology" (2014).

32 As discussed in chapter 5, parents of intersex children tend to prefer DSD nomenclature.

33 See, for example, Vickie Pasterski et al. (2010).

34 This shift from collective confrontation to contested collaboration doesn't necessarily mean that the intersex rights movement is no longer an example of an embodied

health movement. Rather, it suggests that as an embodied health movement, the movement's center has shifted with some organizations willing to work with the medical profession to promote change in intersex medical care.

35 See "Our Mission" a.

36 Ibid.

37 See "Dear ISNA Friends and Supporters."

38 Ibid.

39 See "Our Mission" b.

40 See Hinkle and Viloria (2012).

41 Ibid.

42 Scholars have offered different interpretations of disorder of sex development terminology. Some (e.g., Topp 2013; Davidson 2009; Holmes 2009; Reis 2009, 2007; Karkazis 2008) have openly criticized it. Historian Elizabeth Reis (2009, 2007) has, for example, offered *divergence of sex development*. Communication scholar Sarah Topp (2013) has supported *differences of sex development*. Philosopher Ellen Feder (2009b) has argued that "the change should be understood as normalizing in a positive sense" (134). She has also more directly stated that *DSD* "could be understood as progressive" (2009a, 226). Similarly, academics Alice Dreger and April Herndon (2009) acknowledge that "[r]eception of the new terminology has been mixed among people with intersex" (212), yet they embrace the possibility that the DSD terminology can bring positive change to the community. Sociologist Alyson Spurgas warns us that "the DSD/intersex debate and its associated contest over treatment protocol has consequences for embodied (and thus sexed, gendered and desiring) individuals everywhere" (2009, 118).

CHAPTER 3. MEDICAL JURISDICTION AND THE INTERSEX BODY

1 See Ieuan Hughes (2010a,b) and Vickie Pasterski et al. (2010).

2 A handful of medical experts didn't respond to my invitation to participate, even when I sent several invitations. However, many others welcomed me into their offices. One provider even invited me into their home.

3 Along with Suzanne Kessler and Wendy McKenna (1978), scholars such as Anne Fausto-Sterling (2000a,b, 1993) and Alice Dreger (1998a,b,c) were several of the very few early feminist critics of the medical management of intersex traits. See also Kessler (1998, 1990).

4 See Sarah Richardson (2013).

5 For an in-depth, and balanced, discussion of John Money's career and intellectual contributions, see *Fuckology* (Downing, Morland, and Sullivan 2015). See also David Rubin (2012).

6 Although John Money and Anke Ehrhardt (1972) maintained that the fetal brain's hormonal exposure was correlated with gendered behavior, which gender scholars such as Anne Fausto-Sterling (1985) and Sandra Lipsitz Bem (1993) criticized, socialization was still noted as the significant factor in the formation of one's gender identity. David Rubin (2012) has closely examined John Money's research and has,

in turn, argued "intersex was integral to the historical emergence of the category *gender* as distinct from *sex* in the mid-twentieth-century English-speaking world" (883–84).

7 For an additional critique, see Sandra Lipsitz Bem's *The Lenses of Gender* (1993).

8 For a discussion of twentieth-century intersex medical management in the United States, see Elizabeth Reis's (2009) *Bodies in Doubt*—especially chapters 4 and 5.

9 See John Money and Patricia Tucker (1975); Joseph Zubin and John Money (1973); and John Money and Anke Ehrhardt (1972).

10 Diamond and Sigmundson's 1997 article "Sex Reassignment at Birth: A Long Term Review and Clinical Implications" sparked the media's interest in exposing Money as a research fraud.

11 Sharon Preves describes the impact Diamond and Sigmundson (1997) had in fueling a media frenzy over the John/Joan case in both her book *Intersex & Identity* (2003) and article "Sexing the Intersexed" (2001). For example, the story was covered not only in *Rolling Stone* but also by the *New York Times*, *Newsweek*, *Time*, and NBC's *Dateline* as well as in various other public outlets.

12 See Holmes (1996) for a full transcription of this talk.

13 I first applied Conrad and Schneider's five-stage model of medicalized deviance to the renaming of intersex in 2011 (see Davis 2011) and then again in 2014 (see Davis 2014b). My analysis here is slightly different, as it is based on further research into the nomenclature shift from *intersex* to *DSD*.

14 See Finlayson (2014).

15 Until the medical profession acknowledges and classifies a particular symptom or trait of the body as "abnormal," the medical trait does not officially exist (Conrad 2007; Lorber and Moore 2002; Scott 1990; Zola 1986, 1972; Blaxter 1978).

16 See, for example, Rola Nakhal et al. (2013), Rebecca Deans et al. (2012), J. Pleskacova et al. (2010), Leendert H.J. Looijenga et al. (2007), Martine Cools et al. (2006), and Deborah Merke and Stefan Bornstein (2005).

17 Gayle Rubin (1975) disentangles sex from gender in her theoretical piece "The Traffic in Women: Notes on the 'Political Economy' of Sex."

18 For a discussion of biomedicalization processes see Clarke et al. (2003).

19 It should be noted that there are very few such long-term outcome studies (see Lee and Houk 2012 and Schober et al. 2012).

20 Recall Dr. C.'s comment that "surprisingly, within our setting, very rarely, a psychiatrist" is included in the gender assignment process.

21 See Maarten Hajer and Hendrik Wagenaar (2003) and Rebecca Hannagan and Christopher Larimer (2010).

22 See Jürg C. Streuli et al. (2013).

23 The medical arena is not the only space where testosterone has been used to classify bodies into sex categories. In recent years, for instance, sports governing bodies such as the International Olympic Committee and the International Association of Athletics Federation have created new eligibility rules concerning testosterone that

prohibit women with naturally high testosterone levels from competing as women in elite competitions (see Karkazis et al. 2012).

24 Dr. H. is the rare individual who has been on both sides of medical management: He offers care to individuals with intersexuality, and he himself has an intersex trait.

25 See Ieuan Hughes (2010a,b) and Vickie Pasterski et al. (2010).

26 DSD language is also used by medical experts on intersex traits from other parts of the world. However, given my methodological scope, I focus on the United States.

27 See Barry A. Kogan et al. (2012).

CHAPTER 4. THE POWER IN A NAME

1 The risk of osteoporosis may be increased after surgical modification of intersex bodies (see, e.g., Bertelloni, Baroncelli, and Mora 2010; Soule, Conway, Prelevic, Prentice, Ginsburg, and Jacobs 1995).

2 I first explored these tensions concerning nomenclature in an article titled "The Power in a Name: Diagnostic Terminology and Diverse Experiences" (Davis 2014b).

3 Nikolas Rose and Carlos Novas (2005) build on Adriana Petryna's (2002) conceptualization of biological citizenship.

4 For more information about the individual level of gender structure, see Risman (2004, 1998).

5 Robert Sparrow (2013) has gone so far as to suggest that pre-implantation genetic diagnosis (PGD) prior to in vitro fertilization is morally permissible, when selecting against intersex traits. I have challenged this claim on the grounds that it relies on misinformed assumptions about the visibility of the intersex community and the boundaries it challenges, the origins and status of intersex shame and stigma, and the health risks associated with intersex traits (Davis 2013).

6 In the early 1900s, alleged biological differences between males and females were discussed in various outlets by various sources. See, for example, William Blair Bell (1916); Frank Lillie (1939); Walter Heape (1913); and Walter Young, Robert Goy, and Charles Phoenix (1964). These narrow ideologies about sex differences continue to be supported (e.g., Bakker 2014).

7 In November 2013, Germany began allowing a third gender-indeterminate option on birth certificates. This has been heavily criticized by intersex activists. See, for example, Viloria (2013). In recent years, a few countries, for example Australia and India, have allowed citizens to identify as a third gender on their passports.

8 See, for example, Viloria (2015), Costello (2015a,b), Rebecca Jordan-Young (2010), and Nelly Oudshoorn (1994). For progressive perspectives from medical professionals, see Dr. Lih-Mei Liao et al. (2012).

9 Even medical research reports dissatisfaction with genital surgery and sexual life. See, for example, Birgit Kohler et al. (2012), Verena Schönbucher et al. (2010), and/or Minto et al. (2003).

10 In a review piece, Dr. Sarah Creighton and colleagues (2012) conclude, "Regardless of sex of rearing, many individuals with DSD and atypical genitals will, at some

point, require genital reconstructive surgery" (608). They do, however, acknowledge that "there is ongoing debate as to both the timing of surgery and which procedure should be chosen" (Creighton et al. 2012, 608).

11 Participant has indicated preference for this spelling of the chosen pseudonym. In an earlier publication (Davis 2011), the pseudonym was spelled Pigeon.

12 Even medical professionals acknowledge that their surgical interventions aren't working. See, for example, Birgit Kohler et al. (2012), Verena Schönbucher et al. (2010), Johannsen et al. (2010), and/or Minto et al. (2003). Yet, they continue to perform these interventions. Dr. Jennifer Yang and colleagues (2007) even go so far as to conclude that "nerve sparing reduction clitoroplasty . . . is a safe and reliable approach to correct the enlarged clitoris" (1600).

13 We know relatively little about the health risks associated with intersex traits. For example, there is no clear consensus on the malignancy risks associated with complete androgen insensitivity syndrom (CAIS). Dr. Rebecca Deans and colleagues analyzed sixty-two studies of malignancy risks associated with CAIS and found that gonadal malignancy ranged from 0 percent to 22 percent (Deans et al. 2012). They conclude, "An accurate estimate for adult malignancy risk is unavailable" (Deans et al. 2012, 894). See also Nakhal et al. (2013), Pleskacova et al. (2010), and Cools et al. (2006). Congenital adrenal hyperplasia (CAH) is another trait referenced in claims that intersex traits pose serious health risks. While some forms of CAH can be life-threatening if left untreated, it is also the case that there is a remarkable "range of severity" associated with CAH, which can be minimized with appropriate medical care (Merke and Bornstein 2005, 2125).

14 I first described the sexual struggles associated with intersex in a piece titled "'Bringing Intersexy Back'? Intersexuals and Sexual Satisfaction" (Davis 2014a).

15 Emily Grabham (2012) argues that surgeries aimed at "normalizing" the intersex body are problematic, not necessarily because they disrupt the natural body but because they "interrupt what Bourdieu would term a sense of corporeal 'immersion into the forming'; an immersion which, in his theory of time as social action, is intimately linked with social power and possibilities" (1).

16 See Viloria (2014).

17 See Briffa (2014).

18 It is important to keep in mind that terminological preferences are not static. For example, I am aware of one intersex person who when interviewed expressed preference for DSD over intersex language. However, several years have passed, and this particular person now embraces intersex terminology. I return to this point in chapter 6.

19 Sociologist Betsy Lucal (1999, 793) illustrates how "gender bending, blending, and passing" can disrupt the traditional gender order (793). Thus, Lucal challenges Lorber's (1994) assertion that gender bending and the like highlights, rather than disrupts, the gender structure.

20 See Candace West and Don Zimmerman (1987) for a discussion of "doing gender."

21 Ibid.

22 It is important to acknowledge that the majority of intersex people do not gender transition. This is not to say that all people are satisfied with their gender assignment. Urologist Paulo Sampaio Furtado and colleagues (2012) suggest that 8.5 to 20 percent of intersex people are affected by gender dysphoria.

23 Vanessa is especially interesting because her views on intersex terminology are now quite different from those she held when I interviewed her in October 2009. I return to this point in chapter 6.

24 See Koyama (2006), which is a transcription of a keynote speech she gave in February 2006 at the University of Vermont's Translating Identity Conference.

25 See, for example, Annemarie Jutel (2011, 2009), Phil Brown (2007, 1995, 1990), Peter Conrad (2007), Jennifer Fishman and Laura Mamo (2002), and Elizabeth Cooksey and Phil Brown (1998).

26 See also David Rosenhan (1973).

27 See Wilbur Scott (1990) and Mildred Blaxter (1978).

28 Ibid.

29 See, for example, Barbara Risman (2004, 1998).

30 See, for example, Annemarie Jutel (2011, 2009), Phil Brown (2007, 1995, 1990), Peter Conrad (2007), Jennifer Fishman and Laura Mamo (2002), and Elizabeth Cooksey and Phil Brown (1998).

CHAPTER 5. A DIFFERENT KIND OF INFORMATION

1 Polly Carmichael and Julie Alderson (2004) also document the ways in which intersex affects the entire family.

2 For a greater discussion of the interactional level of gender structure, see Barbara Risman (2004, 1998). See also Candace West and Don Zimmerman (1987), Harold Garfinkel (1967), and Erving Goffman (1976, 1959).

3 While there are other cases of atypical children's bodies wherein doctors perform surgery that is medically unnecessary, such as microtia (the malformation of the ear), such traits are rarely treated as medical emergencies.

4 For a discussion of decisional regret among parents of children who experienced hypospadias repair, see Lorenzo et al. (2014). This parental theme of decisional regret is described in other work as well, including Ellen Feder's (2014) *Making Sense of Intersex* and Katrina Karkazis's (2008) *Fixing Sex*.

5 See Rola Nakhal et al. (2013), Rebecca Deans et al. (2012), J. Pleskacova et al. (2010), Leendert H.J. Looijenga et al. (2007), Martine Cools et al. (2006), and Deborah Merke and Stefan Bornstein (2005).

6 Although Kristin Zeiler and Anette Wickström (2009) acknowledge that medical professionals tend to treat intersex as an emergency, they still seem to assume no responsibility for the medically unnecessary surgical interventions they perform. They state: "[T]o think of one's child as between male and female can be frustrating and painful" (Zeiler and Wickström 2009, 373). They go on to say, "This can explain parents' eagerness to straighten the 'slantwise' perception through sex assignment and often also through agreeing to surgery on their child, even though

intersexuality itself questions the sex dimorphism that we take for granted" (Zeiler and Wickström 2009, 373).

7 It is important to acknowledge that Stephanie E. Hullmann and colleagues (2011) have suggested that the stress some caregivers of intersex children express may have little, if anything, to do with their children's gender or sexual identity. Instead, such stress might be the result of one's own psychosocial health and/or one's personal parenting self-assessment.

8 For example, Claudia Astorino (2014), an intersex woman and sociocultural scholar, offers smart communication advice for parents of intersex children.

9 See also Boyse et al. (2014).

10 *dsd*families describes itself as "an on-line information and support resource for families with children, teens and young adults who have a DSD. The website provides a service: it brings together user-friendly information on the medical management and decision-making in DSD, with psychological support, and sensitive and practical peer support" (see "*dsd*families who?").

11 Nikolas Rose and Carlos Nova (2005) describe connecting online as digital biocitizenship and learning from online resources as informational biocitizenship. I discuss the role of technology in intersex activism in greater detail in chapter 2. For a complementary discussion, see Preves's 2005 article "Out of the O.R. and into the Streets: Exploring the Impact of Intersex Media Activism" in the *Cardozo Journal of Law & Gender*.

12 Medical experts Amy Wisniewski and colleagues (2012) also acknowledge the important of peer support groups. They dedicate an entire chapter in their book *Disorders of Sex Development: A Guide for Parents and Physicians* to the cause.

13 As president of the AIS-DSD Support Group, I have fundraising as one of my top priorities. I want to make sure we have enough funds to cover anyone who wants to attend our conference but can't do so because of financial difficulties.

14 Many medical professionals recommend surgical interventions to prevent malignancy. However, there is little, if any, empirical evidence to support such logic. Dr. Rola Nakhal (2013) and colleagues reported, "No invasive cancers were found [from a] retrospective review of the testicular MR images and histologic reports from 25 patients with [complete androgen insensitivity syndrome] who chose to retain their testes beyond age 16 years and who were imaged between January 2004 and December 2010" (Nakhal et al. 2013, 153). Overall, reports of the cancer risk associated with intersex traits are likely more of a scare tactic than an empirically supported reality, as medical studies are inconsistent in their estimation of malignancy risk across various intersex traits. See also Rebecca Deans et al. (2012), J. Pleskacova et al. (2010), Leendert H.J. Looijenga et al. (2007), Martine Cools et al. (2006), and Deborah Merke and Stefan Bornstein (2005).

15 Elizabeth Reis suggests in both her book *Bodies in Doubt* (2009) and her 2007 article "Divergence of Disorder? The Politics of Naming Intersex" that parents might resist intersex language because *sex* appears within the word, giving it a

sexual connotation. While the word *sex* does appear in disorder of sex development terminology, it is masked by the DSD abbreviation.

16 Betsy Driver, an intersex person and co-founder of Bodies Like Ours, has criticized the inclusion of *intersex* in the LGBT abbreviation. See Driver (2014).

17 M. L. Cull and Margaret Simmonds (2010) offer a similar plea in their piece "Importance of Support Groups for Intersex (Disorders of Sex Development) Patients, Families, and the Medical Profession."

CHAPTER 6. CONCLUSION

1 Emily Grabham (2007) cautions that "claims to full citizenship based on equal rights or recognition run the risk that the responsibilities imposed on sexual minorities as the 'trade-off' in these circumstances will amount to an extension of disciplinary power, thereby compromising the transgressive potential of alternative sexualities and sexual practices" (37).

2 As of February 2015, many of these terms are being discussed in the nomenclature working group that Dr. Peter Lee convened as part of what is being labeled a larger "global endeavor" on the current state of intersex medical care. I am a member of the nomenclature working group, along with a few other intersex people and medical experts on intersex.

3 In "Against the Quiet Revolution: The Rhetorical Construction of Intersex Individuals as Disordered," Sarah Topp (2013) advocates for differences of sex development.

4 See Swiss National Advisory Commission on Biomedical Ethics (2012).

5 See German Ethics Council (2012).

6 See Feminist Newswire (2013).

7 See Kelly (2013).

8 See Nandi (2013).

9 Hida Viloria, a well-known intersex activist, offered a critique at *Advocate.com*. See Viloria (2013).

10 See Zara (2013).

11 See Goldberg (2014).

12 See Inter/Act (2014a).

13 Ibid.

14 See Kondolojy (2014).

15 See Frazier (2014).

16 See Wall and Davis (2014).

17 See Inter/Act (2014b).

18 See Maza (2014).

19 For the current IOC policy, see "IOC Executive Board." For the current IAAF policy on hyperandrogenism, see "IAAF Medical."

20 See Levy (2009).

21 Ibid.

22 See Padmadeo (2014).

23 See "Let Dutee Run."

24 Outside of the United States, Kenya had a groundbreaking lawsuit of its own that ended victoriously in late 2014. John Chigiti, a Kenyan attorney, successfully argued in court that the client he represented, an intersex child, should be issued a birth certificate despite being denied such legal recognition because of the inability to fit into the narrow male/female classification system. See Chigiti (2014).

25 See "Project Integrity."

26 Ibid.

27 See "Complaint."

28 See "Project Integrity."

29 See Greenfield (2014).

30 See Dreger (2013).

31 See Yan and Sutton (2013).

32 See Hall (2013).

33 See WISTV (2013).

34 The defendants' motion to dismiss the case was denied in August 2013 by the federal court and in April 2014 by the state court. See Zieselman (2014).

35 See Southern Poverty Law Center (2013).

36 See Pagonis (2015).

37 Ibid.

38 See Rola Nakhal et al. (2013), Rebecca Deans et al. (2012), J. Pleskacova et al. (2010), Leendert H.J. Looijenga et al. (2007), Martine Cools et al. (2006), and Deborah Merke and Stefan Bornstein (2005).

39 As noted earlier, the risk of cancer associated with intersex conditions is inconsistent. Thus, this number can't reasonably be empirically verified or rejected.

40 According to the U.S. Census Bureau, 74.8 percent of U.S. households have Internet access at home. See U.S. Census Bureau (2014).

41 See AIS-DSD Support Group.

42 On September 21, 2007, *The Oprah Winfrey Show* aired an episode titled "Growing up Intersex." It featured three intersex people and historian Alice Dreger.

43 See Greenhouse (2013).

44 The original airdate of this episode was September 21, 2007.

45 See Inter/Act (2014c).

46 See Inter/Act (2012).

47 Ibid.

48 For example, see Sharon Preves's (2003) book *Intersex and Identity* or Katrina Karkazis's (2008) book *Fixing Sex*.

REFERENCES

Aaronson, Ian A., and Alistair J. Aaronson. 2010. "How Should We Classify Intersex Disorders?" *Journal of Pediatric Urology* 6: 443–46.

Admin. 2012. "Our Mission." *OII-USA: Intersex in America*. Accessed January 31, 2015. http://oii-usa.org/about/our-intersex-mission/.

Agamben, Giorgio. 2005. *State of Exception*. Translated by Kevin Attell. Chicago: University of Chicago Press.

———. 2000. *Means without End: Notes on Politics*. Translated by Vincenzo Binneti and Cesare Casarino. Minneapolis: University of Minnesota Press.

AIS-DSD Support Group. Accessed January 31, 2015. http://aisdsd.org/.

Astorino, Claudia. 2014."Talking About Intersex." *The Parents Project*. Accessed February 8, 2015. http://parentsproject.tumblr.com/post/94830012767/im-a-parent-of-an-intersex-child-and-dont-know.

Bakker, Julie. 2014. "Sex Differentiation: Organizing Effects of Sex Hormones." In *Gender Dysphoria and Disorders of Sex Development: Progress in Care and Knowledge*, edited by Baudewijntje P.C. Kreukels, Thomas D. Steensma, and Annelou L.C. de Vries, 3–23. Boston: Springer.

Barr, Murray L., and Ewart G. Bertram. 1949. "A Morphological Distinction Between Neurones of the Male and Female, and the Behaviour of the Nucleolar Satellite During Accelerated Nucleoprotein Synthesis." *Nature* 163: 676–77.

Bastien-Charlebois, Janik, and Vincent Guillot. 2013. "Medizinische Widerstände gegenüber der Kritik von Intersex Aktivistinnen." Geschlechternormativität und Effekte für Kindheit und Adoleszenz. Internationaler wissenschaftlicher Kongress: 24–29 September 2012, Luxemburg (traduction, sous presse).

Bell, William Blair. 1916. *The Sex Complex: A Study of the Relationship of the Internal Secretions to the Female Characteristics and Functions in Health and Disease*. London: Baillière, Tindall & Cox.

Bem, Sandra. 1993. *The Lenses of Gender: Transforming the Debate on Sexual Inequality*. New Haven: Yale University Press.

Bernstein, Mary. 1997. "Celebration and Suppression: The Strategic Uses of Identity by the Lesbian and Gay Movement." *American Journal of Sociology* 103(3): 531–65.

Bertelloni, Silvano, G. I. Baroncelli, and S. Mora. 2010. "Bone Health in Disorders of Sex Differentiation." *Sexual Development* 4(4/5): 270–84.

Blackless, Melanie, Anthony Charuvastra, Amanda Derryck, Anne Fausto-Sterling, Karl Lauzanne, and Ellen Lee. 2000. "How Sexually Dimorphic Are We? Review and Synthesis." *American Journal of Human Biology* 12(2): 151–66.

Blaxter, Mildred. 1978. "Diagnosis as Category and Process: The Case of Alcoholism." *Social Science & Medicine* 12: 9–17.

Boyse, Kyla L., Melissa Gardner, Donna J. Marvicsin, and David E. Sandberg. 2014. "'It Was an Overwhelming Thing': Parents' Needs After Infant Diagnosis with Congenital Adrenal Hyperplasia." *Journal of Pediatric Nursing* 29(5): 436–41.

Bradley, Susan J., Gillian D. Oliver, Avinoam B. Chernick, and Kenneth J. Zucker. 1998. "Experiment of Nature: Ablatio Penis at Two Months, Sex Reassignment at Seven Months, and a Psychosexual Follow-Up in Young Adulthood." *Pediatrics* 102(1): e9.

Briffa, Tony. 2014. "Tony Briffa Writes on 'Disorders of Sex Development.'" *OII Australia—Intersex Australia.* Accessed January 31, 2015. http://oii.org.au/26808/tony-briffa-on-dsd/.

Brown, Phil. 2007. *Toxic Exposures: Contested Illnesses and the Environmental Health Movement.* New York: Columbia University Press.

———. 1995. "Naming and Framing: The Social Construction of Diagnosis and Illness." *Journal of Health and Social Behavior* 35: 34–52.

———. 1990. "The Name Game: Toward a Sociology of Diagnosis." *The Journal of Mind and Behavior* 11(3/4): 385–406.

Brown, Phil, Stephen Zavestoski, Sabrina McCormick, Brian Mayer, Rachel Morello-Frosch, and Rebecca Gasior Altman. 2004. "Embodied Health Movements: New Approaches to Social Movements in Health." *Sociology of Health & Illness* 26(1): 50–80.

Butler, Judith. 2004. *Undoing Gender.* New York: Routledge.

———. 1993. *Bodies That Matter: On the Discursive Limits of "Sex."* New York: Routledge.

———. [2006]1990. *Gender Trouble: Feminism and the Subversion of Identity.* New York: Routledge.

Carmack, Adrienne. 2014. *Reclaiming My Birth Rights: A Mother's Wisdom Triumphs Over the Harmful Practices of Her Medical Profession.* United States of America: Adrienne Carmack.

Carmichael, Polly, and Julie Alderson. 2004. "Psychological Care of Children and Adolescents with Disorders of Sexual Differentiation and Determination." In *Paediatric and Adolescent Gynaecology: A Multidisciplinary Approach,* edited by Adam H. Balen et al., 158–78. Cambridge: Cambridge University Press.

Casper, Monica J., and Lisa Jean Moore. 2009. *Missing Bodies: The Politics of Visibility.* New York: New York University Press.

Chase, Cheryl. 1998a. "Hermaphrodites with Attitude: Mapping the Emergence of Intersex Political Activism." *GLQ: A Journal of Lesbian and Gay Studies* 4(2): 189–211.

———. 1998b. "Surgical Progress Is Not the Answer to Intersexuality." *Journal of Clinical Ethics* 9(4): 385–92.

———. 1997. "Making Media: An Intersex Perspective." *Images* Fall: 22–25.

———. 1993. "Intersexual Rights." *The Sciences* 33(4): 3.

Chew-Graham, Carolyn, Carl R. May, and Mark S. Perry. 2002. "Qualitative Research and the Problem of Judgement: Lessons from Interviewing Fellow Professionals." *Family Practice* 19(3): 285–89.

Chigiti, John. 2014. "Kenya: We Need a Legal Framework for Intersex Children." *The Star*. Accessed February 8, 2015. http://allafrica.com/stories/201412100790.html.

Chodorow, Nancy. 1978. *The Reproduction of Mothering: Psychoanalysis and the Sociology of Gender*. Berkeley: University of California Press.

Clarke, Adele E., Janet K. Shim, Laura Mamo, Jennifer Ruth Fosket, and Jennifer R. Fishman. 2003. "Biomedicalization: Technoscientific Transformations of Health, Illness, and U.S. Biomedicine." *American Sociological Review* 68(2): 161–94.

Colapinto, John. 2000. *As Nature Made Him: The Boy Who Was Raised as a Girl*. New York: HarperCollins.

———. 1997. "The True Story of John/Joan." *Rolling Stone*, December 11, 54–73, 92–97.

Committee on Genetics: Section on Endocrinology and Section on Urology. 2000. "Evaluation of the Newborn with Developmental Anomalies of the External Genitalia." *Pediatrics* 106(1): 138–42.

"Complaint." "Complaint in *M.C. v. Aaronson* (federal court)." *Advocates for Informed Choice*. Accessed January 31, 2015. http://aiclegal.org/programs/project-integrity/.

Connell, R. 1987. *Gender and Power: Society, the Person, and Sexual Politics*. Stanford, Calif.: Stanford University Press.

Conrad, Peter. 2007. *The Medicalization of Society: On the Transformation of Human Conditions into Treatable Disorders*. Baltimore: The Johns Hopkins University Press.

———. 1992. "Medicalization and Social Control." *Annual Review of Sociology* 18: 209–32.

Conrad, Peter, and Joseph W. Schneider. [1980] 1992. *Deviance and Medicalization: From Badness to Sickness*. St. Louis: The C. V. Mosby Company.

Consortium on the Management of Disorders of Sex Development. 2006a. *Clinical Guidelines for the Management of Disorders of Sex Development in Childhood*. Accessed January 29, 2014. http://www.accordalliance.org/wp-content/uploads/2013/07/clinical.pdf.

———. 2006b. *Handbook for Parents*. Accessed January 29, 2014. http://www.accordalliance.org/wp-content/uploads/2013/07/parents.pdf.

Cooksey, Elizabeth C., and Phil Brown. 1998. "Spinning on Its Axes: DSM and the Social Construction of Psychiatric Diagnosis." *International Journal of Health Services* 28(3): 525–54.

Cooley, Charles Horton. 1902. *Human Nature and the Social Order*. New York: Scribner.

Cools, Martine, and Stenvert L.S. Drop, Katja P. Wolffenbuttel, J. Wolter Oosterhuis, and Leendert H.J. Looijenga. 2006. "Germ Cell Tumors in the Intersex Gonad: Old Paths, New Directions, Moving Frontiers." *Endocrine Reviews* 27(5): 468–84.

Costello, Cary Gabriel. 2015a. *The Intersex Roadshow*. Accessed January 31, 2015. http://intersexroadshow.blogspot.com/.

———. 2015b. *TransFusion*. Accessed January 31, 2015. http://trans-fusion.blogspot.com/.

Crawley, Sara L. 2012. "Autoethnography as Feminist Self-Interview." In *The Sage Handbook of Interview Research: The Complexity of Craft*, 2nd edition, edited by Jaber F. Gubrium, James A. Holstein, Amir B. Marvasti, and Karyn D. McKinney, 143–59. Thousand Oaks, Calif.: Sage.

———. 2008. "The Clothes Make the Trans: Region and Geography in Experiences of the Body." *Journal of Lesbian Studies* 12(4): 365–79.

Creighton, Sarah, Steven D. Chernausek, Rodrigo Romao, Philip Ransley, and Joao Pippi Salle. 2012. "Timing and Nature of Reconstructive Surgery for Disorders of Sex Development—Introduction." *Journal of Pediatric Urology* 8(6): 602–10.

Crissman, Halley P., Lauren Warner, Melissa Gardner, Meagan Carr, Aileen Schast, Alexandra L. Quittner, Barry Kogan, and David E. Sandberg. 2011. "Children with Disorders of Sex Development: A Qualitative Study of Early Parental Experience." *International Journal of Pediatric Endocrinology*, 10: 1–11.

Cull, M. L., and Margaret Simmonds. 2010. "Importance of Support Groups for Intersex (Disorders of Sex Development) Patients, Families, and the Medical Profession." *Sexual Development* 4(4/5): 310–12.

Davidson, Robert J. 2009. "DSD Debates: Social Movement Organizations' Framing Disputes Surrounding the Term 'Disorders of Sex Development.'" *Liminalis—a Journal for Sex/Gender Emancipation and Resistance*. Accessed February 8, 2015. http://www.liminalis.de/2009_03/Artikel_Essay/Liminalis-2009-Davidson.pdf.

Davis, Georgiann. 2014a. "'Bringing Intersexy Back'? Intersexuals and Sexual Satisfaction." In *Sex Matters: The Sexualities and Society Reader*, 4th ed., 11–21. W. W. Norton & Company.

———. 2014b. "The Power in a Name: Diagnostic Terminology and Diverse Experiences." *Psychology & Sexuality* 5(1): 15–27.

———. 2013. "The Social Costs of Preempting Intersex Traits." *The American Journal of Bioethics* 13(10): 51–53.

———. 2011. "'DSD Is a Perfectly Fine Term': Reasserting Medical Authority Through a Shift in Intersex Terminology." In *Sociology of Diagnosis*, edited by PJ McGann and David J. Hutson, 155–82. Wagon Lane, Bingley UK: Emerald.

Davis, Georgiann, and Erin L. Murphy. 2013. "Intersex Bodies as States of Exception: An Empirical Explanation for Unnecessary Surgical Modification." *Feminist Formations* 25(2): 129–52.

Dayner, Jennifer E., Peter A. Lee, and Christopher P. Houk. 2004. "Medical Treatment of Intersex: Parental Perspectives." *The Journal of Urology* 172(4): 1762–65.

Deans, R., S. M. Creighton, L. M. Liao, and G. S. Conway. 2012. "Timing of Gonadectomy in Adult Women with Complete Androgen Insensitivity Syndrome (CAIS): Patient Preferences and Clinical Evidence." *Clinical Endocrinology* 76(6): 894–98.

"Dear ISNA Friends and Supporters." *Intersex Society of North America*. Accessed January 31, 2015. http://www.isna.org/.

Diamond, Milton. 1982. "Sexual Identity, Monozygotic Twins Reared in Discordant Sex Roles and a BBC Follow-Up." *Archives of Sexual Behavior* 11(2): 181–86.

———. 1979. "Sexual Identity and Sex Roles." In *The Frontiers of Sex Research*, edited by Vern L. Bullough, 33–56. Buffalo, N.Y.: Prometheus Books.

———. 1978. "Sexual Identity and Sex Roles." *Humanist* March/April: 16–19.

———. 1976. "Human Sexual Development: Biological Foundation for Social Development." In *Human Sexuality in Four Perspectives*, edited by Frank A. Beach, 22–61. Baltimore: The Johns Hopkins University Press.

———. 1968. "Genetic-Endocrine Interactions and Human Psychosexuality." In *Perspectives in Reproduction and Sexual Behavior*, edited by Milton Diamond. Bloomington: University of Indiana Press.

———. 1965. "A Critical Evaluation of the Ontogeny of Human Sexual Behavior." *The Quarterly Review of Biology* 40(2): 147–75.

Diamond, Milton, and Jameson Garland. 2014. "Evidence Regarding Cosmetic and Medically Unnecessary Surgery on Infants." *Journal of Pediatric Urology* 10: 2–7.

Diamond, Milton, and H. Keith Sigmundson. 1997. "Sex Reassignment at Birth: A Long Term Review and Clinical Implications." *Archives of Pediatric and Adolescent Medicine* 151(3): 298–304.

Distinguished Scientific Award for the Applications of Psychology. 1986. *American Psychologist* 41(4): 354–62.

Downing, Lisa, Iain Morland, and Nikki Sullivan. 2015. *Fuckology: Critical Essays on John Money's Diagnostic Concepts.* Chicago: University of Chicago Press.

Dreger, Alice D., 2013. "When to Do Surgery on a Child With 'Both' Genitalia." *The Atlantic.* Accessed January 31, 2015. http://www.theatlantic.com/health/archive/2013/05/when-to-do-surgery-on-a-child-with-both-genitalia/275884/.

———. 2007. "Why 'Disorders of Sex Development'? (On Language and Life)." *Alicedreger.com.* Accessed February 6, 2015. http://alicedreger.com/dsd.

———. ed. 1999. *Intersex in the Age of Ethics.* Hagerstown, Md.: University Publishing Group, Inc.

———. 1998a. *Hermaphrodites and the Medical Intervention of Sex.* Cambridge, Mass.: Harvard University Press.

———. 1998b. "Ambiguous Sex—or Ambivalent Medicine? Ethical Issues in the Treatment of Intersexuality." *Hastings Center Report* 28(3): 24–35.

———. 1998c. "A History of Intersexuality: From the Age of Gonads to the Age of Consent." *The Journal of Clinical Ethics* 9(4): 345–55.

Dreger, Alice D., Cheryl Chase, Aron Sousa, Philip A. Gruppuso, and Joel Frader. 2005. "Changing the Nomenclature/Taxonomy for Intersex: A Scientific and Clinical Rationale." *Journal of Pediatric Endocrinology & Metabolism* 18: 729–33.

Dreger, Alice D., and April M. Herndon. 2009. "Progress and Politics in the Intersex Rights Movement: Feminist Theory in Action." *GLQ: A Journal of Lesbian and Gay Studies* 15(2): 199–224.

Driver, Betsy. 2014. "Betsy Driver (BLO) on Adding an 'I' to LGBT." *Newwws.* Accessed January 31, 2015. http://vimeo.com/90768808.

"*dsd*families who?" *dsd*families. Accessed January 31, 2015. http://www.dsdfamilies.org/dsdfamilies/index.php.

"DSD Terminology." 2014. *AIS Support Group (UK).* Accessed January 31, 2015. http://www.aissg.org/DEBATES/DSD.HTM.

Dugan, Kimberly B. 2008. "Just Like You: The Dimensions of Identity Deployment in an Antigay Contested Context." In *Identity Work: Negotiating Sameness and Difference in Activist Environments*, edited by Jo Reger, Rachel Einwohner, and Daniel J. Myers, 21–46. Minneapolis: University of Minnesota Press.

Epstein, Julia. 1990. "Either/or—Neither/Both: Sexual Ambiguity and the Ideology of Gender." *Genders* 7 (Spring): 99–142.

Epstein, Steven. 1996. *Impure Science: AIDS, Activism, and the Politics of Knowledge*. Berkeley: University of California Press.

"FAQ." 2008. *Organisation Intersex International*. Accessed December 23, 2008. http://www.intersexualite.org/Organisation_Intersex_International.html.

Fausto-Sterling, Anne. 2000a. *Sexing the Body: Gender Politics and the Construction of Sexuality*. New York: Basic Books.

———. 2000b. "The Five Sexes, Revisited." *The Sciences* 40(4): 18–23.

———. 1993. "The Five Sexes: Why Male and Female Are Not Enough." *The Sciences* 33(2): 20–25.

———. 1985. *Myths of Gender: Biological Theories about Women and Men*. New York: Basic Books.

Feder, Ellen K. 2014. *Making Sense of Intersex: Changing Ethical Perspectives in Biomedicine*. Bloomington: Indiana University Press.

———. 2009a. "Imperatives of Normality: From 'Intersex' to 'Disorders of Sex Development.'" *GLQ: A Journal of Lesbian and Gay Studies* 15(2): 225–47.

———. 2009b. "Normalizing Medicine: Between 'Intersexuals' and Individuals with 'Disorder of Sex Development.'" *Health Care Analysis: An International Journal of Health Care Philosophy and Policy* (17)2: 134–43.

Feinberg, Leslie. 1996. *Transgender Warriors: Making History from Joan of Arc to Dennis Rodman*. Boston: Beacon Press.

Feminist Newswire. 2013. "UN Condemns 'Normalization' Surgeries of Intersex Children." *Feminist Majority Foundation Blog*. Accessed January 31, 2015. http://feminist.org/blog/index.php/2013/02/08/un-condemns-normalization-surgeries-of-intersex-children/.

Ferree, Myra Marx, Beth Hess, and Judith Lorber. 1999. *Revisioning Gender*. Thousand Oaks, Calif.: Sage.

Finlayson, Courtney. 2014. "Lessons from a Christmas Carol: Acceptance and Kindness for Children with Differences of Sex Development." *Huffington Post*. Accessed February 6, 2015. http://www.huffingtonpost.com/courtney-finlayson/lessons-from-a-christmas-_b_6372938.html.

Fischer-Homberger, Esther. 1970. "Eighteenth-Century Nosology and Its Survivors." *Medical History* 14(4): 397–403.

Fishman, Jennifer R., and Laura Mamo. 2002. "What's in a Disorder? A Cultural Analysis of Medical and Pharmaceutical Constructions of Male and Female Sexual Dysfunction." *Women & Therapy* 24(1–2): 179–93.

Foucault, Michel. 1980. "Introduction." In Herculine Barbin, *Herculine Barbin: Being the Recently Discovered Memoirs of a Nineteenth-Century French Hermaphrodite*. Translated by Richard McDougall, vii–xvii. New York: Pantheon.

———. 1978. *The History of Sexuality, Volume 1: An Introduction*. Translated by Robert Hurley. New York: Vintage.

Frazier, Ran Aubrey. 2014. "Fox News Hosts Put Down Facebook's New Gender Options." *Advocate.com*. Accessed January 31, 2015. http://www.advocate.com/politics/media/2014/02/17/watch-fox-news-hosts-put-down-facebooks-new-gender-options.

Friedman, Asia. 2013. *Blind to Sameness: Sexpectations and the Social Construction of Male and Female Bodies*. Chicago: University of Chicago Press.

Furtado, Paulo Sampaio, Felipe Moraes, Renata Lago, Luciana Oliveira Barros, Maria Betânia Toralles, and Ubirajara Barroso Jr. 2012. "Gender Dysphoria Associated with Disorders of Sex Development." *Nature Reviews Urology* 9(11): 620–27.

Gadpaille, Warren J. 1980. "Biological Factors in the Development of Human Sexual Identity." *Psychiatric Clinics of North America* 3(1): 3–20.

Gamson, Joshua. 1995. "Must Identity Movements Self-Destruct? A Queer Dilemma." *Social Problems* 42(3): 390–407.

Garfinkel, Harold. 1967. *Studies in Ethnomethodology*. Englewood Cliffs, N.J.: Prentice-Hall.

Gearhart, John P. 1996. Interviewed in Natalie Angier's "Intersexual Healing: An Anomaly Finds a Group." *New York Times*, 4 February.

German Ethics Council. 2012. "Press Release: Intersex People Should Be Recognized, Supported and Protected from Discrimination." *Deutscher Ethikrat*. Accessed January 31, 2015. http://www.ethikrat.org/files/press-release-2012-01.pdf/.

Ghaziani, Amin. 2008. *The Dividends of Dissent: How Conflict and Culture Work in Lesbian and Gay Marches on Washington*. Chicago: University of Chicago Press.

Giddens, Anthony. 1990. *The Consequences of Modernity*. Stanford, Calif.: Stanford University Press.

———. 1984. *The Constitution of Society: Outline of the Theory of Structuration*. Berkeley: University of California Press.

Goffman, Erving. 1976. "Gender Display." *Studies in the Anthropology of Visual Communication* 3: 69–77.

———. 1959. *The Presentation of Self in Everyday Life*. New York: Anchor Books.

Goldberg, Lesley. 2014. "MTV's 'Faking It' to Tell Intersex Story in Season 2." *The Hollywood Reporter*. Accessed January 31, 2015. http://www.hollywoodreporter.com/live-feed/mtvs-faking-tell-intersex-story-732076.

Gough, Brendan, Nicky Weyman, Julie Alderson, Gary Butler, and Mandy Stoner. 2008. "'They Did Not Have a Word': The Parental Quest to Locate a 'True Sex' for Their Intersex Children." *Psychology & Health* 23(4): 493–507.

Grabham, Emily. 2012. "Bodily Integrity and the Surgical Management of Intersex." *Body & Society* 18(2): 1–26.

———. 2007. "Citizen Bodies, Intersex Citizenship." *Sexualities* 10(1): 29–48.

Greenberg, Julie A. 2012. *Intersexuality and the Law: Why Sex Matters*. New York: New York University Press.

Greenfield, Charlotte. 2014. "Should We 'Fix' Intersex Children?" *The Atlantic*. Accessed January 31, 2015. http://www.theatlantic.com/health/archive/2014/07/should-we-fix-intersex-children/373536/.

Greenhouse, Emily. 2013. "A New Era for Intersex Rights." *The New Yorker*. Accessed January 31, 2015. http://www.newyorker.com/news/news-desk/a-new-era-for-intersex-rights.

Groves, Robert M., Floyd J. Fowler Jr., Mick P. Couper, James M. Lepkowski, Eleanor Singer, and Roger Tourangeau. 2004. *Survey Methodology*. New York: Wiley.

Hagerman, Margaret A. 2010. "'I Like Being Intervieeeeeeewed! Kids' Perspectives on Participating in Social Research." In *Children and Youth Speak for Themselves*, edited by Heather B. Johnson, 61–105. Wagon Lane, Bingley, UK: Emerald.

Hajer, Maarten A., and Hendrik Wagenaar, eds. 2003. *Deliberative Policy Analysis: Understanding Governance in the Network Society*. Cambridge: Cambridge University Press.

Hall, Lee. 2013. "A Case of Sexual Reassignment Without Consent: Justice for M.C." *CounterPunch*. Accessed January 31, 2015. http://www.counterpunch.org/2013/06/19/justice-for-m-c/.

Hannagan, Rebecca J., and Christopher W. Larimer. 2010. "Does Gender Composition Affect Group Decision Outcomes? Evidence from a Laboratory Experiment." *Political Behavior* 32(1): 51–67.

Harding, Sandra. 1998. *Is Science Multicultural? Postcolonialisms, Feminisms, and Epistemologies*. Bloomington: Indiana University Press.

Hartsock, Nancy. 1983. "The Feminist Standpoint: Developing the Ground for a Specifically Historical Feminist Materialism." In *Discovering Reality: Feminist Perspectives on Epistemology, Metaphysics, Methodology, and Philosophy of Science*, edited by Sandra Harding and Merrill Hintikka, 283–310. Amsterdam: D. Reidel Publishing Company.

Heape, Walter. 1913. *Sex Antagonism*. London: Constable.

Hester, J. David. 2004. "Intersexes and the End of Gender: Corporeal Ethics and Post-gender Bodies." *Journal of Gender Studies* 13(3): 215–25.

Hinkle, Curtis E., and Hida Viloria. 2012. "Ten Misconceptions about Intersex." *OII-USA: Intersex in America*. January 17, 2012. http://oii-usa.org/1144/ten-misconceptions-intersex/.

Hird, Myra. 2000. "Gender's Nature: Intersexuality, Transsexualism and the 'Sex'/'Gender' Binary." *Feminist Theory* 1(3): 347–64.

"History of the Society." *Pediatric Endocrine Society*. Accessed January 31, 2015. https://www.pedsendo.org/about/history/index.cfm.

Holmes, Morgan, ed. 2009. *Critical Intersex*. Surrey, England: Ashgate Publishing.

———. 2008. *Intersex: A Perilous Difference*. Selinsgrove, Pa.: Susquehanna University Press.

———. 1996. "Is Growing up in Silence Better Than Growing up Different?" *Intersex Society of North America*. Accessed January 31, 2015. http://www.isna.org/node/743.

hooks, bell. 1994. *Teaching to Transgress: Education as the Practice of Freedom*. New York: Routledge.

Houk, Christopher P., Ieuan A. Hughes, S. Faisal Ahmed, Peter A. Lee, and Writing Committee for the International Intersex Consensus Conference Participants. 2006. "Summary of Consensus Statement on Intersex Disorders and their Management." *Pediatrics* 118(2): 753–57.

Hughes, Ieuan A. 2010a. "The Quiet Revolution: Disorders of Sex Development." *Best Practice & Research Clinical Endocrinology & Metabolism* 24(2): 159–62.

———. 2010b. "How Should We Classify Intersex Disorders?" *Journal of Pediatric Urology* 6: 447–48.

Hughes, Ieuan A., Christopher Houk, S. Faisal Ahmed, Peter A. Lee, and LWPES1/ESPE2 Consensus Group. 2006. "Consensus Statement on Management of Intersex Disorders." *Archives of Disease in Childhood* 91(7): 554–63.

Hullmann, Stephanie E., David A. Fedele, Cortney Wolfe-Christensen, Larry L. Mullins, and Amy B. Wisniewski. 2011. "Differences in Adjustment by Child Developmental Stage Among Caregivers of Children with Disorders of Sex Development." *International Journal of Pediatric Endocrinology* 16: 1–7.

Hunt, Scott A., and Robert D. Benford. 1994. "Identity Talk in the Peace and Justice Movement." *Journal of Contemporary Ethnography* 22(4): 488–517.

IAAF Medical. "Hyperandrogenism and Sex Reassignment." *International Association of Athletics Federations.* Accessed January 31, 2015. http://www.iaaf.org/about-iaaf/documents/medical.

Imperato-McGinley, Julianne, Ralph E. Peterson, Teofilo Gautier, and Erasmo Sturia. 1979. "Androgens and Evolution of Male-Gender Identity Among Male Pseudohermaphrodites with 5α-Reductase Deficiency." *New England Journal of Medicine* 300(22): 1233–37.

Inter/Act. 2014a. "What Is Intersex? An Intersex FAQ by Inter/Act." *Inter/Act Youth.* Accessed January 31, 2015. http://interactyouth.org/post/97343969730/want-to-support-us-check-out-our.

———. 2014b. "Facebook's Gender Options: An Open Letter to Fox News." *GLAAD.* Accessed January 31, 2015. http://www.glaad.org/blog/guest-post-facebooks-gender-options-open-letter-fox-news.

———. 2014c. *Advocates for Informed Choice.* Accessed January 31, 2015. http://aiclegal.org/programs/interact/.

———. 2012. "We Feel It Is Time We Told Our Own Stories and Spoke Our Own Truths." *Advocates for Informed Choice.* Accessed January 31, 2015. https://aiclegal.files.wordpress.com/2012/10/interact_ms-updated.pdf.

IOC Executive Board. "IOC Regulations on Female Hyperandrogenism." *International Olympic Committee.* Accessed January 31, 2015. http://www.olympic.org/Documents/Commissions_PDFfiles/Medical_commission/IOC-Regulations-on-Female-Hyperandrogenism.pdf.

Johannsen, T. H., C. P. L. Ripa, E. Carlsen, J. Starup, O. H. Nielsen, M. Schwartz, K. T. Drzewiecki, E. L. Mortensen, and K. M. Main. 2010. "Long-Term Gynecological Outcomes in Women with Congenital Adrenal Hyperplasia Due to 21-Hydroxylase Deficiency." *International Journal of Pediatric Endocrinology,* Article ID 784297: 1–7.

Jordan-Young, Rebecca M. 2010. *Brain Storm: The Flaws in the Science of Sex Differences*. Cambridge, Mass.: Harvard University Press.

Jordan-Young, Rebecca M., Peter H. Sönksen, and Katrina Karkazis. 2014. "Sex, Health, and Athletes." *British Medical Journal* 348: g2926.

Jorgensen, Christine. [1967] 2000. *Christine Jorgensen: A Personal Autobiography*. San Francisco: Cleis Press.

Jutel, Annemarie. 2011. *Putting a Name to It: Diagnosis in Contemporary Society*. Baltimore: The Johns Hopkins University Press.

———. 2009. "Sociology of Diagnosis: A Preliminary Review." *Sociology of Health & Illness* 31(2): 278–99.

Karkazis, Katrina. 2008. *Fixing Sex: Intersex, Medical Authority, and Lived Experience*. Durham, N.C.: Duke University Press.

Karkazis, Katrina, Rebecca M. Jordan-Young, Georgiann Davis, and Silvia Camporesi. 2012. "Out of Bounds? A Critique of the New Policies on Hyperandrogenism in Elite Female Athletes." *The American Journal of Bioethics* 12(7): 3–16.

Katzenstein, Mary Fainsod. 1990. "Feminism within American Institutions: Unobtrusive Mobilization in the 1980s." *Signs: Journal of Women in Culture and Society* 16(1): 27–54.

Kelly, Ashlee. 2013. "New Jersey Senate Panel Approves Trans and Intersex Birth Certificate Bill." *Gay Star News*. Accessed January 31, 2015. http://www.gaystarnews.com/article/new-jersey-senate-panel-approves-trans-and-intersex-birth-certificate-bill141213.

Kessler, Suzanne J. 1998. *Lessons from the Intersexed*. New Brunswick, N.J.: Rutgers University Press.

———. 1990. "The Medical Construction of Gender: Case Management of Intersexed Infants." *Signs: Journal of Women in Culture and Society* 16(1): 3–26.

Kessler, Suzanne J., and Wendy McKenna. 1978. *Gender: An Ethnomethodological Approach*. Chicago: University of Chicago Press.

Kitzinger, Celia. 2005. "Heteronormativity in Action: Reproducing the Heterosexual Nuclear Family in After-hours Medical Calls." *Social Problems* 52(4): 477–98.

Kogan, Barry A., Melissa Gardner, Adrianne N. Alpern, Laura M. Cohen, Mary Beth Grimley, Alexandra L. Quittner, and David E. Sandberg. 2012. "Challenges of Disorders of Sex Development: Diverse Perceptions Across Stakeholders." *Hormone Research in Paediatrics* 78(1): 40–46.

Kohler, Birgit, Eva Kleinemeier, Anke Lux, Olaf Hiort, Annette Gruters, Ute Thyen, and the DSD Network Working Group. 2012. "Satisfaction with Genital Surgery and Sexual Life of Adults with XY Disorders of Sex Development: Results from the German Clinical Evaluation Study." *Journal of Clinical Endocrinology & Metabolism* 97(2): 577–88.

Kondolojy, Amanda. 2014. "Tuesday Cable Ratings: 'Sons of Anarchy' Tops Night + '19 Kids and Counting,' 'Little People, Big World,' 'Ink Master,' & More." *TV by the Numbers*. Accessed January 31, 2015. http://tvbythenumbers.zap2it.com/2014/09/24/tuesday-cable-ratings-sons-of-anarchy-tops-night-19-kids-and-counting-little-people-big-world-ink-master-more/306516/.

Koyama, Emi. 2006. "From 'Intersex' to 'DSD': Toward a Queer Disability Politics of Gender." *Intersex Initiative.* Accessed January 31, 2015. http://www.intersexinitiative. org/articles/intersextodsd.html.

Koyama, Emi, and Lisa Weasel. 2002. "From Social Construction to Social Justice: Transforming How We Teach About Intersexuality." *Women's Studies Quarterly* 30(3/4): 169–78.

Lee, Peter A., and Christopher P. Houk. 2013. "Key Discussions from the Working Party on Disorders of Sex Development (DSD) Evaluation, Foundation Merieux, Annecy, France, March 14–17, 2012." *International Journal of Pediatric Endocrinology* 12: 1–8.

———. 2012. "Long-Term Outcomes and Adjustment Among Patients with DSD Born with Testicular Differentiation and Masculinized External Genital Genitalia." *Pediatric Endocrinology Reviews* 10(1): 140–51.

———. 2010. "Review of Outcome Information in 46,XX Patients with Congenital Adrenal Hyperplasia Assigned/Reared Male: What Does It Say About Gender Assignment?" *International Journal of Pediatric Endocrinology.* Article ID 982025: 1–7.

Lee, Peter A., Christopher P. Houk, S. Faisal Ahmed, and Ieuan A. Hughes. 2006. "Consensus Statement on Management of Intersex Disorders." *Pediatrics* 118(2): 488–500.

Lee, Peter A., Justine Schober, Anna Nordenström, Piet Hoebeke, Christopher Houk, Leendert Looijenga, Gianantonio Manzoni, William Reiner, Christopher Woodhouse. 2012. "Review of Recent Outcome Data of Disorders of Sex Development (DSD): Emphasis on Surgical and Sexual Outcomes." *Journal of Pediatric Urology* 8(6): 611–15.

Lee, Peter A., Amy Wisniewski, Laurence Baskin, Maria G. Vogiatzi, Eric Vilain, Stephen Rosenthal, and Christopher Houk. 2014. "Advances in Diagnosis and Care of Persons with DSD Over the Last Decade." *International Journal of Pediatric Endocrinology* 19: 1–13.

"Let Dutee Run." Accessed January 31, 2015. www.letduteerun.org.

Levy, Ariel. 2009. "Either/Or." *The New Yorker.* Accessed January 31, 2015. http://www. newyorker.com/magazine/2009/11/30/eitheror.

Liao, Lih-Mei, Laura Audi, Ellie Magritte, Heino F.L. Bahlburg, and Charmian A. Quigley. 2012. "Determinant Factors of Gender Identity: A Commentary." *Journal of Pediatric Urology* 8(6): 597–601.

Lillie, Frank R. 1939. "Biological Introduction." In *Sex and Internal Secretions,* 2nd ed., edited by E. Allen. Baltimore: Williams & Wilkins.

Looijenga, Leendert H.J., Remko Hersmus, J. Wolter Oosterhuis, Martine Cools, Stenvert L.S. Drop, and Katja P. Wolffenbuttel. 2007. "Tumor Risk in Disorders of Sex Development (DSD)." *Best Practice & Research Clinical Endocrinology & Metabolism* 21(3): 480–95.

Lorber, Judith. 1994. *Paradoxes of Gender.* New Haven: Yale University Press.

Lorber, Judith, and Lisa Jean Moore. 2002. *Gender and the Social Construction of Illness.* Walnut Creek, Calif.: Rowman-Littlefield/Alta Mira Press.

Lorde, Audre. 1984. *Sister Outsider: Essays & Speeches by Audre Lorde*. Berkeley, Calif.: Crossing Press.

Lorenzo, Armando J., João L. Pippi Salle, Bozana Zlateska, Martin A. Koyle, Darius J. Bägli, and Luis H. P. Braga. 2014. "Decisional Regret After Distal Hypospadias Repair: Single Institution Prospective Analysis of Factors Associated with Subsequent Parental Remorse or Distress." *The Journal of Urology* 191(5): 1558–63.

Lucal, Betsy. 1999. "What It Means to Be Gendered Me: Life on the Boundaries of a Dichotomous Gender System." *Gender & Society* 13(6): 781–97.

MacKenzie, Gordene Olga. 1994. *Transgender Nation*. Bowling Green, Ohio: Bowling Green State University Press.

Magritte, Ellie. 2012. "Working Together in Placing the Long Term Interests of the Child at the Heart of the DSD Evaluation." *Journal of Pediatric Urology* 8(6): 571–75.

Mak, Geertje. 2012. *Doubting Sex: Inscriptions, Bodies and Selves in Nineteenth-Century Hermaphrodite Case Histories*. Manchester, UK: Manchester University Press.

Malouf, Matthew A., and Arlene Baratz. 2012. "Disorders or Differences of Sex Development." In *Improving Emotional & Behavioral Outcomes for LGBT Youth: A Guide for Professionals*, edited by Sylvia K. Fisher, Jeffrey M. Poirier, and Gary M. Blau, 67–86. Baltimore: Paul H. Brookes Publishing Co.

Martin, Patricia Yancey. 2005. *Rape Work: Victims, Gender and Emotions in Organization and Community Context*. New York: Routledge/Taylor & Francis Group.

———. 2004. "Gender as Social Institution." *Social Forces* 82(4): 1249–73.

Maza, Carlos. 2014. "Watch This Fox News Host's Heartfelt Apology to the Intersex Community." *Media Matters for America*. Accessed January 31, 2015. http://mediamatters.org/blog/2014/03/01/watch-this-fox-news-hosts-heartfelt-apology-to/198300.

Meadow, Tey. 2011. "'Deep Down Where the Music Plays': How Parents Account for Childhood Gender Variance." *Sexualities* 14(6): 725–47.

Melucci, Alberto. 1989. *Nomads of the Present: Social Movements and Individual Needs in Contemporary Society*. Philadelphia: Temple University Press.

Merke, Deborah P., and Stefan R. Bornstein. 2005. "Congenital Adrenal Hyperplasia." *The Lancet* 365(9477): 2125–36.

Meyer-Bahlburg, Heino. 1996. Interviewed in Natalie Angier's "Intersexual Healing: An Anomaly Finds a Group." *New York Times*, 4 Februrary.

Meyerowitz, Joanne. 2002. *How Sex Changed: A History of Transsexuality in the United States*. Cambridge, Mass.: Harvard University Press.

Minto, Catherine L., Lih-Mei Liao, Christopher R.J. Woodhouse, Phillip G. Ransley, and Sarah M. Creighton. 2003. "The Effect of Clitoral Surgery on Sexual Outcome in Individuals Who Have Intersex Conditions with Ambiguous Genitalia: A Cross-Sectional Study." *The Lancet* 361(9365): 1252–57.

Money, John, and Anke A. Ehrhardt. 1972. *Man & Woman, Boy & Girl: Differentiation and Dimorphism of Gender Identity from Conception to Maturity*. Baltimore: The Johns Hopkins University Press.

Money, John, and Patricia Tucker. 1975. *Sexual Signatures on Being a Man or a Woman*. Boston: Little, Brown.

Money, John, Joan G. Hampson, and John L. Hampson. 1957. "Imprinting and the Establishment of Gender Role." *Archives of Neurology and Psychiatry* 77(3): 333–36.

Moore, Keith L., and Murray L. Barr. 1955. "Smears from the Oral Mucosa in the Detection of Chromosomal Sex." *The Lancet* 269(6880): 57–58.

Moore, Keith L., M. A. Graham, and Murray L. Barr. 1953. "The Detection of Chromosomal Sex in Hermaphrodites from a Skin Biopsy." *Surgery, Gynecology, and Obstetrics* 96(6): 641–48.

Morland, Iain. 2011. "Intersex Treatment and the Promise of Trauma." In *Gender and the Science of Difference: Cultural Politics of Contemporary Science and Medicine*, edited by Jill A. Fisher, 147–63. New Brunswick, N.J.: Rutgers University Press.

———. 2009. "Between Critique and Reform: Ways of Reading the Intersex Controversy." In *Critical Intersex*, edited by Morgan Holmes, 191–213. Surrey, England: Ashgate Publishing.

Morris, Sherri G. 2006. "Twisted Lies: My Journey in an Imperfect Body." In *Surgically Shaping Children: Technology, Ethics, and the Pursuit of Normality*, edited by Erik Parens, 3–12. Baltimore: The Johns Hopkins University Press.

Moshiri, Mariam, Teresa Chapman, Patricia Y. Fechner, Theodore J. Dubinsky, Margarett Shnorhavorian, Sherif Osman, Puneet Bhargava, and Douglas S. Katz. 2012. "Evaluation and Management of Disorders of Sex Development: Multidisciplinary Approach to a Complex Diagnosis." *RadioGraphics* 32(6): 1599–1618.

Nakhal, Rola S., Margaret Hall-Craggs, Alex Freeman, Alex Kirkham, Gerard S. Conway, Rupali Arora, Christopher R. J. Woodhouse, Dan N. Wood, and Sarah M. Creighton. 2013. "Evaluation of Retained Testes in Adolescent Girls and Women with Complete Androgen Insensitivity Syndrome." *Radiology* 268(1): 153–60.

Nandi, Jacinta. 2013. "Germany Got It Right by Offering a Third Gender Option on Birth Certificates." *The Guardian*. Accessed January 31, 2015. http://www.theguardian.com/commentisfree/2013/nov/10/germany-third-gender-birth-certificate.

Naples, Nancy A. 2003. *Feminism and Method: Ethnography, Discourse Analysis, and Activist Research*. New York: Routledge.

"NOW: Leading the Fight." *National Organization for Women*. Accessed January 31, 2015. http://now.org/resource/now-leading-the-fight/.

Oakley, Ann. 1981. "Interviewing Women: A Contradiction in Terms." In *Doing Feminist Research*, edited by Helen Roberts, 30–62. London: Routledge & Kegan Paul.

Oudshoorn, Nelly. 1994. *Beyond the Natural Body: An Archeology of Sex Hormones*. New York: Routledge.

"Our Mission" a. *Intersex Society of North America*. Accessed January 31, 2015. http://www.isna.org/.

"Our Mission" b. *Accord Alliance*. Accessed January 31, 2015. http://www.accordalliance.org/about-accord-alliance/our-mission/.

Padmadeo, Vinayak Bhushan. 2014. "Dutee Chand Knocks on CAS Door." *The Indian Express*. Accessed January 31, 2015. http://indianexpress.com/article/sports/sport-others/dutee-knocks-on-cas-door/.

Pagonis, Pidgeon. 2015. "Update on M.C.'s Case—The Road to Justice Can Be Long, But There Is More Than One Path for M.C." *Advocates for Informed Choice*. Accessed January 31, 2015. http://aiclegal.org/update-on-the-m-c-case-the-road-to-justice-can-be-long/.

Pasterski, Vickie, Kiki Mastroyannopoulou, Deborah Wright, Kenneth J. Zucker, and Ieuan A. Hughes. 2014. "Predictors of Posttraumatic Stress in Parents of Children Diagnosed with a Disorder of Sex Development." *Archives of Sexual Behavior* 43(2): 369–75.

Pasterski, Vickie, P. Prentice, and I. A. Hughes. 2010. "Impact of the Consensus Statement and the New DSD Classification System." *Best Practice & Research Clinical Endocrinology & Metabolism* 24(2): 187–95.

Peggy. 2014. "Woozle Hunting," *Bodies Like Ours: Intersex Information and Peer Support*. Accessed January 31, 2015. http://www.bodieslikeours.org/forums/showthread.php?p=28667.

———. 2013. "Mutating Misinformation on Intersex Prevalence." *Bodies Like Ours: Intersex Information and Peer Support*. Accessed January 31, 2015. http://www.bodieslikeours.org/forums/showthread.php?p=28475.

Petryna, Adriana. 2003. *Life Exposed Biological Citizens After Chernobyl*. Princeton, N.J.: Princeton University Press.

Pfeffer, Carla A. 2014. "'I Don't Like Passing as a Straight Woman': Queer Negotiations of Identity and Social Group Membership." *American Journal of Sociology* 120(1): 1–44.

Pleskacova, J., R. Hersmus, J. W. Oosterhuis, B. A. Setyawati, S. M. Faradz, M. Cools, K. P. Wolffenbuttel, J. Lebl, S. L. Drop, and L. H. Looijenga. 2010. "Tumor Risk in Disorders of Sex Development." *Sexual Development* 4(4/5): 259–69.

Polletta, Francesca, and James M. Jasper. 2001. "Collective Identity and Social Movements." *Annual Review of Sociology* 27: 283–305.

Preves, Sharon. 2005. "Out of the O.R. and into the Streets: Exploring the Impact of Intersex Media Activism." *Cardozo Journal of Law & Gender* 12(1): 247–88. (Reprinted from *Research in Political Sociology*.)

———. 2003. *Intersex and Identity: The Contested Self*. New Brunswick, N.J.: Rutgers University Press.

———. 2001. "Sexing the Intersexed: An Analysis of Sociocultural Responses to Intersexuality." *Signs: Journal of Women in Culture and Society* 27(2): 523–56.

———. 2000. "Negotiating the Constraints of Gender Binarism: Intersexuals' Challenge to Gender Categorization." *Current Sociology* 48(3): 27–50.

"Project Integrity." *Advocates for Informed Choice*. Accessed January 31, 2015. http://aiclegal.org/programs/project-integrity/.

Reger, Jo, Daniel J. Myers, and Rachel L. Einwohner, ed. 2008. *Identity Work in Social Movements*. Minneapolis: University of Minnesota Press.

Reis, Elizabeth. 2009. *Bodies in Doubt: An American History of Intersex*. Baltimore: The Johns Hopkins University Press.

———. 2007. "Divergence or Disorder? The Politics of Naming Intersex." *Perspectives in Biology and Medicine* 50(4): 535–43.

Richardson, Sarah S. 2013. *Sex Itself: The Search for Male and Female in the Human Genome*. Chicago: University of Chicago Press.

Risman, Barbara. 2004. "Gender as a Social Structure: Theory Wrestling with Activism." *Gender & Society* 18(4): 429–50.

———. 1998. *Gender Vertigo: American Families in Transition*. New Haven: Yale University Press.

Rose, Nikolas. 2007. *The Politics of Life Itself: Biomedicine, Power, and Subjectivity in the Twenty-First Century*. Princeton, N.J.: Princeton University Press.

———. 2001. "The Politics of Life Itself." *Theory, Culture & Society* 18(6): 1–30.

Rose, Nikolas, and Carlos Novas. 2005. "Biological Citizenship." In *Global Assemblages: Technology, Politics, and Ethics as Antrhopological Problems*, edited by Aihwa Ong and Stephen J. Collier, 439–63. Malden, Mass.: Blackwell Publishing.

Rosenhan, David. L. 1973. "On Being Sane in Insane Places." *Science* 179(4070): 250–58.

Rubin, David A. 2012. "'An Unnamed Blank That Craved a Name': A Genealogy of Intersex as Gender." *Signs: Journal of Women in Culture and Society* 37(4): 883–908.

Rubin, Gayle. 1975. "The Traffic in Women: Notes on the 'Political Economy' of Sex." In *Toward an Anthropology of Women*, edited by Rayna R. Reiter, 157–210. New York: Monthly Review Press.

Rupp, Leila J., and Verta Taylor. 2003. *Draq Queens at the 801 Cabaret*. Chicago: University of Chicago Press.

———. 1999. "Forging Feminist Identity in an International Movement: A Collective Identity Approach to Feminism." *Signs: Journal of Women in Culture and Society* 24(2): 363–86.

Sandberg, David E., and Tom Mazur. 2014. "A Noncategorical Approach to the Psychosocial Care of Persons with DSD and Their Families." In *Gender Dysphoria and Disorders of Sex Development: Progress in Care and Knowledge*, edited by Baudewijntje P.C. Kreukels, Thomas D. Steensma, and Annelou L.C. de Vries, 93–114. Boston: Springer.

Sanders, Caroline, Bernie Carter, and Lynne Goodacre. 2012. "Parents Need to Protect: Influences, Risks, and Tensions for Parents of Prepubertal Children Born with Ambiguous Genitalia." *Journal of Clinical Nursing* 21(21/22): 3315–23.

Schilt, Kristen, and Laurel Westbrook. 2009. "Doing Gender, Doing Heteronormativity: 'Gender Normals,' Transgender People, and the Social Maintenance of Heterosexuality." *Gender & Society* 23(4): 440–64.

Schmitt, Frederika E., and Patricia Yancey Martin. 1999. "Unobtrusive Mobilization by an Institutionalized Rape Crisis Center: 'All We Do Comes from Victims.'" *Gender & Society* 13(3): 364–84.

Schober, Justine, Anna Nordenstrom, Piet Hoebeke, Peter Lee, Christopher Houk, Leendert Looijenga, Gianantonio Manzoni, William Reiner, and Christopher

Woodhouse. 2012. "Disorders of Sex Development: Summaries of Long-Term Outcome Studies." *Journal of Pediatric Urology* 8(6): 616–23.

Schönbucher, Verena, Katinka Schweizer, and Hertha Richter-Appelt. 2010. "Sexual
Quality of Life of Individuals with Disorders of Sex Development and a 46,XY
Karyotype: A Review of International Research." *Journal of Sex & Marital Therapy*
36(3): 193–215.

Scott, Wilbur J. 1990. "PTSD in DSM-III: A Case in the Politics of Diagnosis and Disease." *Social Problems* 37(3): 294–310.

Serano, Julia. 2007. *Whipping Girl: A Transsexual Woman on Sexism and the Scapegoat-
ing of Femininity*. Emeryville, Calif.: Seal Press.

Slijper, F. M. E., P. G. Frets, A. L. M. Boehmer, S. L. S. Drop, and M. F. Niermeijer.
2000. "Androgen Insensitivity Syndrome (AIS): Emotional Reactions of Parents
and Adult Patients to the Clinical Diagnosis of AIS and Its Confirmation by Androgen Receptor Gene Mutation Analysis." *Hormone Research in Paediatrics* 53(1): 9–15.

Soule, S. G., G. Conway, G. M. Prelevic, M. Prentice, J. Ginsburg, and H. S. Jacobs.
1995. "Osteopenia as a Feature of the Androgen Insensitivity Syndrome." *Clinical
Endocrinology* 43(6): 671–75.

Southern Poverty Law Center. 2013. "Sex-Assignment Surgery on a Child Ruled Un-
constitutional." *Salon*. Accessed January 31, 2015. http://www.salon.com/2013/08/24/
judge_deems_sex_assignment_surgery_on_a_child_violation_of_constitution_
partner/.

Sparrow, Robert. 2013. "Gender Eugenics? The Ethics of PGD for Intersex Conditions."
American Journal of Bioethics 13(10): 29–38.

Sprague, Joey. 2005. *Feminist Methodologies for Critical Researchers: Bridging Differences*. Lanham, Md.: Rowman & Littlefield.

Spurgas, Alyson K. 2009. "(Un)Queering Identity: The Biosocial Production of Inter-
sex/DSD." In *Critical Intersex*, edited by Morgan Holmes, 97–122. Surrey, England:
Ashgate Publishing.

Still, Brian. 2008. *Online Intersex Communities: Virtual Neighborhoods of Support and
Activism*. Amherst, N.Y.: Cambria Press.

Streuli, Jürg C., Effy Vayena, Yvonne Cavicchia-Balmer, and Johannes Huber. 2013.
"Shaping Parents: Impact of Contrasting Professional Counseling on Parents'
Decision Making for Children with Disorders of Sex Development." *The Journal of
Sexual Medicine* 10(8): 1953–60.

Stryker, Susan. 2008. *Transgender History*. Berkeley, Calif.: Seal Press.

Swiss National Advisory Commission on Biomedical Ethics. 2012. "On the Management of Differences of Sex Development: Ethical Issues Relating to 'Intersexuality.'"
Swiss National Advisory Commission on Biomedical Ethics. Accessed January 31,
2015. http://www.nek-cne.ch/fileadmin/nek-cne-dateien/Themen/Stellungnahmen/
en/NEK_Intersexualitaet_En.pdf.

Tamar-Mattis, Anne. 2006. "Exceptions to the Rule: Curing the Law's Failure to Protect
Intersex Infants." *Berkeley Journal of Gender, Law, & Justice* 21(4): 59–110.

Tamar-Mattis, Anne, Arlene Baratz, Katharine Baratz Dalke, and Katrina Karkazis. 2014. "Emotionally and Cognitively Informed Consent for Clinical Care for Differences of Sex Development." *Psychology & Sexuality* 5(1): 44–55.

Taylor, Verta. 2000. "Mobilizing for Change in a Social Movement Society." *Contemporary Sociology* 29(1): 219–30.

———. 1999. "Gender and Social Movements: Gender Processes in Women's Self-Help Movements." *Gender & Society* 13(1): 8–33.

———. 1996. *Rock-a-by Baby: Feminism, Self-Help, and Postpartum Depression*. New York: Routledge.

———. 1995. "Self-Labeling and Women's Mental Health: Postpartum Illness and the Reconstruction of Motherhood." *Sociological Focus* 28(1): 23–47.

———. 1989. "Social Movement Continuity: The Women's Movement in Abeyance." *American Sociological Review* 54(5): 761–75.

Taylor, Verta, and Nancy Whittier. 1992. "Collective Identity in Social Movement Communities: Lesbian Feminist Mobilization." In *Frontiers in Social Movement Theory*, edited by Aldon Morris and Carol Mueller, 104–29. New Haven: Yale University Press.

Topp, Sarah S. 2013. "Against the Quiet Revolution: The Rhetorical Construction of Intersex Individuals as Disordered." *Sexualities* 16(1/2): 180–94.

Turner, Stephanie S. 1999. "Intersex Identities: Locating New Intersections of Sex and Gender." *Gender & Society* 13(4): 457–79.

U.S. Census Bureau. 2014. "Computer & Internet Trends in America." *United States Census Bureau: Measuring America*. Accessed January 31, 2015. http://www.census.gov/hhes/computer/files/2012/Computer_Use_Infographic_FINAL.pdf.

Valocchi, Steve. 1999. "Riding the Crest of a Protest Wave? Collective Action Frames in the Gay Liberation Movement, 1969–1973." *Mobilization: An International Quarterly* 4(1): 59–73.

Veith, Ilza. 1969. "Historical Reflections on the Changing Concepts of Disease." *California Medicine* 110(6): 501–6.

Viloria, Hida. 2015. *Hida Viloria: Intersex Writer and Activist*. Accessed January 31, 2015. http://hidaviloria.com/.

———. 2014. "What's in a Name: Intersex and Identity." *Advocate.com*. Accessed January 31, 2015. http://www.advocate.com/commentary/2014/05/14/op-ed-whats-name-intersex-and-identity.

———. 2013. "Germany's Third-Gender Law Fails on Equality." *Advocate.com*. Accessed January 31, 2015. http://www.advocate.com/commentary/2013/11/06/op-ed-germany%E2%80%99s-third-gender-law-fails-equality.

Wall, Sean Saifa, and Georgiann Davis. 2014. "Hey, Fox News, Intersex Is Not a Punchline." *Advocate.com*. Accessed January 31, 2015. http://www.advocate.com/commentary/2014/02/26/op-ed-hey-fox-news-intersex-not-punchline.

Warren, Carol AB. 2014. "Gender Reassignment Surgery in the 18th Century: A Case Study." *Sexualities* 17(7): 872–84.

West, Candace, and Don H. Zimmerman. 1987. "Doing Gender." *Gender & Society* 1(2): 125–51.

Whyte, William Foote. 1984. *Learning from the Field: A Guide from Experience*. Newbury Park, Calif.: Sage.

Wisniewski, Amy B., Steven D. Chernausek, and Bradley P. Kropp. 2012. *Disorders of Sex Development: A Guide for Parents and Physicians*. Baltimore: The Johns Hopkins University Press.

WISTV. 2013. "Man Speaks Out on Being Born Both Sexes." *WISTV. com*. Accessed January 31, 2015. http://www.wistv.com/story/22431745/man-speaks-out-on-being-born-both-sexes.

Yan, Holly, and Joe Sutton. 2013. "Parents Sue South Carolina for Surgically Making Child Female." *CNN*. Accessed January 31, 2015. http://www.cnn.com/2013/05/15/health/child-sex-surgery-suit/.

Yang, Jennifer, Diane Felsen, and Dix P. Poppas. 2007. "Nerve Sparing Ventral Clitoroplasty: Analysis of Clitoral Sensitivity and Viability." *The Journal of Urology* 178(4): 1598–1601.

Young, William C., Robert W. Goy, and Charles H. Phoenix. 1964. "Hormones and Sexual Behavior." *Science* 143(3603): 212–18.

Zara, Christopher. 2013. "Intersex Australia: Third Gender Allowed on Personal Documents in Addition to Male and Female." *International Business Times*. Accessed January 31, 2015. http://www.ibtimes.com/intersex-australia-third-gender-allowed-personal-documents-addition-male-female-1307843.

Zeiler, Kristin, and Anette Wickström. 2009. "Why Do 'We' Perform Surgery on Newborn Intersexed Children? The Phenomenology of the Parental Experience of Having a Child with Intersex Anatomies." *Feminist Theory* 10(3): 359–77.

Zieselman, Kimberly. 2014. "South Carolina Court Rejects Attempt to Delay Justice for Child Subjected to Unnecessary Sex-Assignment Surgery." *Advocates for Informed Choice*. Accessed January 31, 2015. http://aiclegal.org/wordpress/wp-content/uploads/2012/10/AIC-Lawsuit-Updates-Press-Release_4.9.14_Final.pdf.

Zola, Irving K. 1986. "Illness Behaviour: A Political Analysis." In *Illness Behavior: A Multidisciplinary Model*, edited by Sean McHugh and T. Michael Vallis, 213–18. New York: Plenum Press.

———. 1972. "Medicine as an Institution of Social Control." *The Sociological Review* 20(4): 487–504.

Zubin, Joseph, and John Money, eds. 1973. *Contemporary Sexual Behavior: Critical Issues in the 1970s*. Baltimore: The Johns Hopkins University Press.

Zuger, Bernard. 1975. "Comments on 'Gender Role Differentiation in Hermaphrodites.'" *Archives of Sexual Behavior* 4(5): 579–81.

———. 1970. "Gender Role Determination: A Critical Review of the Evidence from Hermaphroditism." *Psychosomatic Medicine* 32(5): 449–67.

INDEX

Aaronson, Alistair, 46

Aaronson, Ian, 46

abnormality: feelings of anxiety about, 22, 93–97, 101–102; frameworks for defining, 56; intersex perceived as, 21, 26, 30–31, 68, 72, 88, 90–97, 119–120, 141–142, 146; rejecting idea of intersex as, 96–97, 113

Accord Alliance: collaborative strategies with medical professionals, 48–50, 53; formation of, 48–49; mission and goals, 29, 38–39, 49–50; participant recruitment from, 11, 19

actions for positive social change, 24, 156–168

ADHD. *See* attention deficit hyperactivity disorder

Advocates for Informed Choice (AIC), 148–150, 154–156, 167–168

Agamben, Giorgio, 8, 23, 117–118, 142

AIC. *See* Advocates for Informed Choice

AIS. *See* androgen insensitivity syndrome

AIS-DSD Support Group (formerly AISSG-USA), 34–36; board, 36; collaboration with medical community, 51–52, 53; membership, 5, 34–35, 113; mission, 50; name change, 34, 51, 161; named AIS-DSD Support Group for Women and Families, 34–35; president, 5, 161, 168, 188n13; size and power of, 156; youth programming, 167–168

AISSG-USA (Androgen Insensitivity Syndrome Support Group–USA): annual meeting, 1, 3, 6, 9–10, 26–27, 87,

163–164; mission and goals, 29, 34–36; name change, 34, 51, 161; parent involvement, 128–131; participant recruitment, 11, 19; scholarships for annual conference, 132; support for intersex research, 5. *See also* AIS-DSD Support Group (formerly AISSG-USA)

"All Gender Restroom" sign, 35, 36

American Psychological Association (APA), 59; Distinguished Scientific Award for the Applications of Psychology, 63

American Psychologist, 59

androgen exposure during gestation, 71

androgen insensitivity syndrome (AIS), 2, 34, 91, 95, 1, 64, 182n25

Androgen Insensitivity Syndrome Support Group–USA. *See* AISSG-USA

APA. *See* American Psychological Association

asexuals, 12, 93, 97; terminological preferences, 105–106

As Nature Made Him (Colapinto), 62

Astorino, Claudia, 188n8

attention deficit hyperactivity disorder (ADHD), 114, 181n9

Australia, third gender on government-issued personal documents, 148, 185n7

authority. *See* medical authority

autoethnography, 168

Baratz, Arlene, 48, 130

Bem, Sandra Lipsitz, 183n6

Benjamin, Harry, 30

ABOUT THE AUTHOR

Georgiann Davis is Assistant Professor of Sociology at the University of Nevada, Las Vegas.